T0332753

True Nutrition, True Fitness

True Nutrition,
True Fitness

by

Jerrold Winter, PhD

 **Humana Press**

Totowa, New Jersey

To

Barbara, Jessica, Kurt, Jerry, and Anne

© 1991 The Humana Press Inc.
999 Riverview Drive, Suite 208
Totowa, NJ 07512

Printed in the United States of America. 9 8 7 6 5 4 3 2

Library of Congress Cataloging in Publication Data
Main entry under title:

True nutrition, true fitness / by Jerrold Winter.

416 p. 15.24 x 22.86 cm

Includes bibliographical references (6 pages) and index.
ISBN 0-89603-184-5
1. Nutrition. 2. Health. I. Title.
RA784.W67 1991
613.2—dc20
90-5214
CIP

Preface

This book is an attempt to bring order and genuine understanding to the thousands of bits of information on nutrition, exercise, and their relationships to human health that seem to swirl so unceasingly about us. To accomplish this end, dear reader, we must, as Will Durant suggested many years ago, "put aside our fear of inevitable error" and strive to attain "total perspective."

Several factors render the task of achieving that total perspective more difficult to realize than we might first imagine. In our free society, every citizen enjoys the inalienable right to preach to every other citizen on such matters as sex and religion and gun control. And sometimes it seems that they all do—perhaps especially in the areas of nutrition and health—so that the noise level we shall need to overcome is high. Their, and my, personal freedom to expound on these topics stops only just short of trying to sell you arthritis cures or taking up hammer and chisel to open your skull—licenses are required for these latter activities. And when our near absolute right of free speech is further compounded with the abundant energies of our capitalism, matters often grow more exciting still.

Were I to go to the busiest corner in town and tell passers-by that a macrobiotic diet will cure lung cancer, few would hear me. But take the same message to a publisher and it may be distributed to millions. (Lest you think my example a bit extreme, Chapter 21 discusses a book assuring us that just such a macrobiotic diet will cure lung cancer.) A man who makes his living as a publisher stated the simple truth a few years ago: it is "economically rewarding...to bring out yet another hyperbolic, monomaniacal fad book from the ranks of fearmongers and zealots."

Oddly enough, the very success of modern science can also work against us. Science informs us and science confuses us. A lack of understanding of science leads to the belief that science is magic and scientists magicians. But if science is magic, then could not magic be science? If electrons and viruses, though never seen, are real and influence my life, then why should I not believe that coffee enemas cure cancer, that alien beings regularly visit the earth, and that I can lose fifty pounds while eating all I want? The answer lies in an understanding of the ways of science. Science demands evidence; it is not enough to say that something is true. Faith is a wonderful thing, but it is not science.

That is not to say that science is always correct. Error is inevitable, even in science, though we may act rationally despite it. The lack of absolute certainty of science, especially medical science, is all too often obscured by rote learning methods of our schools and by the finality implicit in a thirty-second science feature on the evening news.

The present book was written for all those who share with me the desire to bring order to our knowledge of nutrition and fitness, and to lift that knowledge to the level of wisdom. If this strikes you as a bit pretentious, the book itself nonetheless falls into that most humble class of writings, the "self-help" text. Yes, the reader will learn about cholesterol levels and aerobic conditioning and vitamin B_{12} injections. But these and the host of other practical topics discussed will always be placed in an historical context. And what we believe today will be presented not as dogma, but as the best current approximation of truth based upon what has come before.

The chapters of the book are arranged in three parts. The first and longest is concerned with nutrition, the second with exercise, and the third brings these elements together as they relate to cancer, cardiovascular disease, osteoporosis, and weight loss. There is a reasonable sequence to each of these parts, but I have with intent made each chapter nearly independent of the others. It is my hope that intelligible, informative, and even pleasurable reading can be found wherever the book is opened. And so I would not object if someone with a particular interest in weight loss were to begin at the very last chapter.

I have heard it said that Giacomo Narroni began a musical composition in 1868 and that 50 years later, on his deathbed, he was still revising it. Lest this book suffer the same fate as Narroni's piece, I have settled for less than the completeness and perfection of treatment my subjects deserve. On the other hand, I am buoyed by the hope that what follows will provide a rational framework for the many things about nutrition left still unwritten and for those yet to be discovered.

Jerrold Winter, PhD

Acknowledgments

There are many to whom I owe thanks. Among those who will remain nameless are the staff of the Health Sciences Library of the State University at Buffalo. The superb holdings of the HSL have provided me many pleasureable hours and form the backbone of this book. The medical literature, including those thousands of reprints in my personal collection, is of course entirely the work of countless scientists, physicians, and scholars; some have been named in this book, many more have not; to each I am indebted.

My colleagues in the Department of Pharmacology and Therapeutics of the School of Medicine and Biomedical Sciences helped me in many ways. In particular I thank Edward A. Carr, Jr., MD, my chairman during most of the time that this book was in progress, for providing an environment conducive to scholarship. My introduction to electronic word and data processing was provided by Alan M. Reynard, PhD; without the tools he first provided me, this work would have been an impossible burden. Although I must retain responsibility for any and all possible errors of fact or interpretation, I wish to thank all those who, in passing through my laboratory, kindly read and commented upon parts of the manuscript. Among those persons are Ms. Susan Regan, Dr. Patrice Ferriola Bruckenstein, Ms. Sistine Chen, Dr. Patricia Noker, Ms. Karen McCann, Dr. David McCann, Ms. Katherine Bonson, and Ms. Deborah Petti.

Finally, I must express my deep gratitude to Thomas Lanigan of the Humana Press.

Contents

PART I

NUTRITION

Introduction

...nutritional literacy....will enable you...to separate fact from fiction, proof from mere plausibility, quackery from honest endeavor, and thus to make sensible decisions about your own daily diet.

Human nutrition is a remarkably complex subject. In a standard textbook of nutrition that sits on my shelf, 81 authors manage to fill nearly 1400 pages, yet fall far short of being all inclusive. My treatment of the subject is somewhat shorter. Nonetheless, I believe that the information contained in the following pages will take you a long way toward nutritional literacy. This literacy will enable you, in matters of nutrition, to separate fact from fiction, proof from mere plausibility, quackery from honest endeavor, and thus to make sensible decisions about your own daily diet.

We might best begin by covering the macronutrients, which are only three in number: protein, fat, and carbohydrate. With the micronutrients—all those substances we consume in only small quantities—we will be less comprehensive. For example, of the 13 known vitamins, 12 are discussed; vitamin K is not included because it is used primarily in a medical setting and is rarely promoted to the general public. Of the dozens of minerals and trace metals having a known or likely place in nutrition, I have focused chiefly upon calcium and iron. These were chosen because the human requirement for each is controversial, because each is related to a major disease, and because dietary supplements of each are sold in enormous quantities.

An understanding of the roots of human nutrition rests on the basic medical sciences of biochemistry and physiology, but its branches touch and are in turn influenced by factors far removed from science.

The controversy presently surrounding the National Research Council's well-known Recommended Dietary Allowances offers an excellent example.

The Food and Nutrition Board, established in 1941, first issued its Recommended Dietary Allowances (RDAs) in 1943, and since has published updated versions at roughly five-year intervals. Throughout this book, I will make frequent reference to the most recent set of RDAs, the Tenth Edition, which appeared in late 1989, nearly five years behind schedule. Work on the Tenth Edition had actually begun in 1980. Funding for the project came from a $600,000 contract between the National Institutes of Health (NIH) and the National Research Council, parent organization of the Food and Nutrition Board.

But on October 7, 1985, Dr. Frank Press, Chairman of the National Research Council, informed the Director of the NIH that the Tenth Edition of the RDAs would not be issued "at this time." He stated that this decision was based upon differences of opinion between the RDA Revision Committee and members of the Food and Nutrition Board. Subsequently, a subcommittee of the Food and Nutrition Board itself was appointed to complete work on the Tenth Edition. No members of the original Revision Committee were included.

The precise reasons for the rejection of the Revision Committee's report have never been stated officially. It is likely that the decision was based primarily upon the recommendation by the committee that the RDA for vitamins A and C be reduced. Those reductions appeared to some to be in direct conflict with the National Cancer Institute's campaign to increase consumption of foods rich in vitamins A and C. For the interested reader, the chapters on vitamins A, C, and B_{12}, folic acid, and iron include references to publications in 1987 by individual members of the 1980 Revision Committee.

CHAPTER 1

Protein

Too Much of a Good Thing?

...many continue to think of protein as "energy food"....in fact protein is essential in the human diet only as a source of amino acids. If no protein were ever used for energy, we would be none the worse for it.

Protein, fat, and carbohydrate are the three major components of the diet. They are sometimes called macronutrients to distinguish them from the vitamins and minerals, which are needed in much smaller quantities. Of the three macronutrients, only protein has an unblemished reputation among the general public. After all, protein is the substance of muscle and most of us yearn for more shapely or more massive muscles. Protein supplements are advertized in every body-building and weightlifting magazine. Children and adults alike are incessantly told they need protein by the milk producers, the cattlemen, and other self-serving groups.

Popular notions about protein are as likely to be erroneous as are those of just about any other aspect of nutrition. This is not a new phenomenon. Around 1850, Baron Justus von Liebig, a famous and influential German chemist, stated

> *Physically, all proteins are structured like strands
> of pearls, in which each pearl corresponds to an
> amino acid....It is the individual amino acid, not
> the protein as a whole, that is absorbed by the gut.
> This...makes nonsense of the often-heard claim that
> eating larger quantities of a particular protein will
> cause more of that protein to appear in the body.*

that protein is the sole source of energy for the contraction of muscle. Despite von Liebig's lack of evidence, and despite a series of experiments in the second half of the 19th century that disproved the idea, many continue to think of protein as "energy food." Though it is true that the body can use protein for energy if fat and carbohydrate are lacking, in fact protein is essential in the human diet only as a source of amino acids. If no protein were ever used for energy, we would be none the worse for it.

Physically, all proteins are structured like strands of pearls, in which each pearl corresponds to an amino acid and the string that connects one to another is a chemical (peptide) bond. When we eat the muscle of an animal, whether it be the breast of a turkey, the rump of a steer, or the leg of a lamb, we cannot make direct use of the proteins they contain. Instead, our digestive enzymes (which, by the way, are themselves proteins) break the chemical bonds connecting the protein pearls and release their individual beads of amino acids. It is the individual amino acid, not the protein as a whole, that is absorbed from the gut. This fact of digestive biology makes nonsense of the often-heard claim that eating larger quantities of a particular protein will cause more of that protein to appear in the body. A current example of this erroneous claim is often made by the peddlers of the enzyme, superoxide dismutase, which is claimed to extend life. In truth, my gut cares not at all about what it may be told in my local health food store; superoxide dismutase will be digested like any other protein. To be sure, my body will simultaneously manufacture superoxide dismutase, as human bodies have done for a million years or so, but the quantity will be completely independent of how much of the commercial product I may have eaten. A similar fantasy—more or less the inverse of the first—is that eating large amounts of a single amino acid found in a particular protein will promote synthesis of that protein.

Tryptophan is an amino acid that has long been popular in health food and vitamin stores. Advertized as "nature's tranquilizer" and as a "natural sleeping pill" to be used "instead of drugs," tryptophan nicely illustrates the hazards of equating "natural" with "harmless" or, for that matter, of equating "natural" with "truly natural."

Beginning in the fall of 1989, reports began to come into the Centers for Disease Control in Atlanta of a peculiar illness in people who had taken tryptophan. The majority of cases were women, possibly a reflection of the recent promotion of tryptophan for treatment of premenstrual syndrome (PMS) and bulemia. The illness was given the name "eosinophilia–myalgia syndrome." Eosinophils are a kind of blood cell that increases in number in response to parasitic infection. Myalgia refers to the muscle pain that is also characteristic of the syndrome.

In November 1989, the Food and Drug Administration stopped the sale of tryptophan. As of February 1990, the Centers for Disease Control had collected 1046 cases of eosinophilia-myalgia syndrome including seven deaths. A final point: most tryptophan sold in this country is manufactured in Japan using bacterial ferments containing *E. coli,* a principal bacterium of the human colon. One thought is that the eosinophilia-myalgia syndrome is not caused by tryptophan, but by a chemical or biological contaminant introduced during the manufacturing process.

*Beef is a high quality protein...it provides
the essential amino acids in proportions...close to what
the body needs. However, beef is not necessarily a good
choice as a primary source of protein for humans.*

Once it was recognized that all proteins are made up of amino acids, nutritionists could begin to ask intelligent questions about the relative importance of each of the basic amino acids. By 1910, four essential amino acids had been identified. No matter what quantities of the other amino acids were included in the diet, test animals would not thrive unless these four were present. It is now known that of the 20 or so different amino acids that make up all of the various proteins of the human body, only about half are essential. So, if we wished to design an ideal dietary protein, that protein would provide all of the essential amino acids in the exact proportions required by the body, and nothing more. The degree to which actual proteins conform to this ideal is commonly referred to as the protein's "quality."

"High quality protein" has a nice sound to it and, given a choice, we would certainly be inclined to choose high quality over low. But this is not always a good idea. The reason is that most measures of protein quality are based on the ability of a protein to produce a high rate of growth in young rats or farm animals. In planning a diet for a human being with a life expectancy of perhaps 80 years, several factors other than growth rate have to be considered.

Beef is a high quality protein, i.e., it provides the essential amino acids in proportions that are close to what the body needs. However, beef is not necessarily a good choice as a primary source of protein for humans. It is not the beef protein itself that is at fault, but what comes with it. For example, there is reason to believe that the saturated fats found so abundantly in beef contribute not only to the flavor of the

> *Vegetables...provide protein of relatively low "quality," but since [it] is not accompanied by much... saturated fat, they can be said to be "better."*

meat, but to cerebral stroke and heart disease as well. (Much more will be said about this later.) Vegetables, on the other hand, provide protein of relatively low "quality," but since their protein is not accompanied by much in the way of saturated fat, they can be said to be "better."

In choosing a source of protein, we also have to consider the number of calories that it provides. A quart of milk gives us about 80% of our daily protein requirement. If it is whole milk, it will also contribute about 650 calories. On the other hand, a quart of skim milk containing exactly the same amount of protein has only 320 calories. If we are interested in feeding starving people, providing whole milk is a much better idea—because they have need of a source of energy as well as of protein. On the other hand, if we are typical Americans fighting a never ending battle against our waistlines, the difference between a quart of skim and a quart of whole milk each day works out to about half a pound of fat a week.

Millions of people throughout the world are protein deficient. The disease that they suffer has been given the Ghanian name, kwashiorkor. Victims of the disease may retain substantial amounts of body fat, yet they are severely malnourished and will soon die if untreated. The appearance of children is especially deceptive. The puffy cheeks, thickened limbs, and potbellies they exhibit are not signs of overfeeding, as many who see photographs of these truly starving bodies so readily assume, but instead signify the abnormal accumulation of fluid in their bodies.

Because the human immune system is dependent upon protein synthesis, bacterial and viral infections of all kinds are a constant threat to the lives of those with kwashiorkor.

...protein deficiency in America is...
often simply the result of a poor choice of foods—
what Laurence Finberg has called
the triumph of ignorance over affluence.

In the United States, protein malnutrition is seen most often in children and in the elderly. In the latter group, the deficiency may lead to progressive confusion, a symptom that has sometimes been interpreted as the irreversible brain deterioration of Alzheimer's disease. Though it is sometimes directly caused by poverty, protein deficiency in America is more often simply the result of a poor choice of foods—what Laurence Finberg has called the triumph of ignorance over affluence. The money paid for an order of french fries at a fast-food counter would provide enough protein for a day if it were used instead to buy milk.

Just how much protein do we actually need? More than 100 years ago, Edward Smith, a British physician and physiologist, estimated the human requirement to be 80–90 grams per day. The figure has moved steadily downward since then; in the United States the estimate of the protein needs of a one-year-old has fallen by more than 50% since 1948. Current recommendations for an adult male of average size range from 30 to 70 grams, depending upon which group of national or international experts one consults. The primary reason for such a wide range among these estimates is that no one has complete confidence in the underlying data. For example, some experimental studies intended to determine the quality of egg protein have come up with values only half those found in others. To allow for such uncertainties, the present Recommended Dietary Allowance (RDA) in the United States is quite generous: 63 grams for men and 50 grams for women. These amounts are based on body weight, for protein of average quality, and allow for variation between individuals.

A recent survey of the American diet found that the average person eats about twice the current RDA for protein. The total intake of protein hasn't changed much in the last 80 years, but there has been a definite trend away from vegetable protein. In 1910, about half of dietary protein came from vegetables, but today more than two-thirds is of animal origin, with beef, pork, and poultry products the major contributors. I don't know of anyone who believes we need all that protein—and certainly not all that animal protein. There simply isn't any evidence that protein in excess is a good idea. The late Conrad Elvehjem, one of America's greatest nutritional scientists, once said that, as far as protein is concerned, minimal and adequate and optimal are all the same thing.

To put an average daily requirement of 60 grams of protein into a more familiar perspective, let's convert it to ounces: 2.1 oz. Thus we actually require only about 2 oz of average quality protein per day, which is equivalent to less than 7 oz of ground beef, chicken, or tunafish. Indeed, in a 1989 report called "Diet and Health," the Food and Nutrition Board of the National Research Council called for a reduction in the amount of animal protein consumed. Even strict vegetarians, those who eat no foods of animal origin, need not risk protein deficiency if a proper variety of plant foods is consumed. Although an individual vegetable protein may be relatively poor in one or more of the essential amino acids, simply eating combinations of vegetables will usually suffice to compensate. And indeed, a number of detailed studies have found that a mixed vegetarian diet can provide adequate amounts of all the essential amino acids.

Conrad Elvehjem, one of America's greatest
nutritional scientists, once said that,
as far as protein is concerned, minimal and
adequate and optimal are all the same thing.

...all of us, vegetarians included, are cannibals.
Each day, for example...I digest about 70 grams
of the high quality protein made by my own body.
It comes from spent digestive enzymes, cells
sloughed from the stomach and intestines, and a
variety of other sources of proteinaceous debris.

The fact that a purely vegetarian diet can deliver adequate protein should not lead us to believe that all "vegetarian" diets are without risk of protein deficiency, however. As with a variety of other nutritional insults, infants and children are most likely to be harmed by excessive reliance on vegetables as protein sources. This was illustrated in 1982 by Drs. Eric Shinwell and Rafael Gorodischer, who treated some of the children born into a strictly vegetarian religious cult in Israel. Of 25 infants brought to their hospital, three were dead on arrival and five others died within hours of admission. Seven others among these children suffered from kwashiorkor. After weaning at about three months of age, no milk had been provided. Instead, the major constituent of the infants' diet was a dilute "milk" made from soybean flour. Despite persistent efforts over a three-year period, Shinwell and Gorodischer were unable to induce the leaders of the group to modify their dietary practices.

It is not necessary for an infant to be born into a religious commune in the Middle East to be at risk of protein deficiency. Beginning in 1983, many parents in this country and in Canada succumbed to advertisements for Edensoy, a soybean-based drink recommended by the manufacturer as a substitute for mother's milk. Despite warnings by the FDA, Edensoy continued to be promoted in "health food stores" as infant formula. The Edensoy story reached its deserved end when, in January 1989, Michael J. Potter, president of

Eden Foods of Clinton, Michigan, gained the distinction of being the first person sentenced to prison for violation of the Infant Formula Act of 1980. Precisely how many infants were actually harmed by Edensoy remains still unknown.

Another oft-repeated fallacy regarding protein is that each meal must contain all of the essential amino acids in just the right proportions. This idea seems to have originated in the realization in the early 1940s that protein synthesis in the body's cells requires the simultaneous presence of all the essential amino acids. Thus, it is reasoned, we must eat a completely balanced array of amino acids each time we consume protein. But there is another source of protein to be considered. The fact is that all of us, vegetarians included, are cannibals. Each day, for example, in addition to the protein in my food, I digest about 70 grams of the high quality protein made by my own body. It comes from spent digestive enzymes, cells sloughed from the stomach and intestines, and a variety of other sources of proteinaceous debris. What reaches my liver is a mix of amino acids derived from both my diet and from the recycling of my very self. Thus, in the short run, the liver is able to pump an optimal mix of essential amino acids into the bloodstream, even when there are large fluctuations in the protein content of individual meals.

With the exception of those who consciously avoid all meat, fish, and dairy products, few of us truly need to give much thought to getting enough protein. Should we, on the other hand, be concerned about too much? The answer to that question is not entirely clear. Certainly we must be aware of the other substances that come with the protein—I've already mentioned the saturated fat from animal sources. In times of famine, the incidence of allergies decreases and some

In times of famine, the incidence of allergies decreases and some suggest that this is a beneficial effect of decreased protein intake.

suggest that this is a beneficial effect of decreased protein intake. The role of protein in the origin of kidney disease is uncertain, but kidneys that are already damaged seem to benefit from a low protein diet. There has also been concern about the production of cancer-causing substances when proteins are cooked at high temperatures.

Summing Up

Though it is true that millions of persons around the world suffer from protein deficiency, this condition is relatively uncommon in the US and other well-developed countries. Surveys indicate that the average American consumes about twice the RDA for protein. Despite what body building magazines may tell us, there is no evidence that protein in excess is a good idea. Indeed, there is at least the suspicion that some of us may be harmed by too much protein. For these reasons, I think it a good (and economical) idea to consciously limit our intake of animal protein. A diet in which the only sources of protein are cereals, vegetables, and skim milk or low-fat yogurt is nearly ideal. For those who cringe at the thought of life without steaks, chops, and roasts, I will simply suggest that you let meat be the crowning glory of your major meals rather than the mainstay of each and every breakfast, lunch, and dinner.

CHAPTER 2

Fat and Essential Fatty Acids

Lessons from Sheep Sex

---

*Fat is the black beast of contemporary America.
From puberty to senility, millions are engaged
in a war against it....Never mind that nearly all
the weapons provided us in this war
against our excess avoirdupois are worthless;
they at least give hope that science will soon let us
eat all we want while staying slim....*

Fat is the black beast of contemporary America. From puberty to senility, millions are engaged in a war against it. Our allies in this war range from unabashed hucksters ("Lose fat while you sleep.") to physicians who practice "bariatric medicine" (irreverently referred to by some as "fat doctors"). Never mind that nearly all the weapons provided us in this war against our excess avoirdupois are worthless; they at least give hope that science will soon let us eat all we want while staying slim as Michael Jackson or one of the models in *Glamour*.

Unless we abruptly return to a Botticellian view of beauty, the efforts of the advertising industry alone are capable of maintaining the war on fat; the Pepsi Generation has no fat people going for the gusto in Marlboro Country. In addition, however, the medical establishment has for years warned that obesity contributes to high blood pressure and

---

*Unless we abruptly return to a Botticellian view
of beauty, the efforts of the advertising industry
alone are capable of maintaining the war on fat...*

---

heart disease and diabetes. More recently, the notion has
been set loose that the fat we eat contributes to certain forms
of cancer. I will defer until Part III a consideration of weight
loss and the possible influence of dietary fat and obesity on
disease. The reason is that none of those complicated issues
can be understood without some notion of what fat is and
why, in fact, certain forms of dietary fat are as essential to
our health as are proteins or any of the vitamins.

Most of us feel that we have a good understanding of
fat. One all too familiar form is subcutaneous fat, the amor-
phous material found beneath our skin. In modest amounts,
it softens and gives beauty to the bony skeleton with its
overlying muscle. In excess, subcutaneous fat gives substance
to terms like potbelly, love handles, and thunder thighs. Few
of us ever see the fat of our bodies without its covering of
skin, but human fat is very little different from that of any
other animal—the white-to-pale yellow edge of a steak, for
instance. Vegetables contain fat as well. Because vegetable
fats tend to be liquid at room temperature, they are called
oils. There are many different fats and oils with many different chemical compositions, but we may describe a typical
fat as a molecule of glycerin to which has been attached three
fatty acids; this combination is called a triglyceride. Each of
the three fatty acids of a triglyceride is a string of carbon
atoms. A saturated fatty acid is one in which each of the carbon atoms has two atoms of hydrogen attached to it. A monounsaturated fatty acid has two fewer than the full comple-
ment of hydrogen atoms and is said to possess one unsaturated bond. Fatty acids with two, three, or more unsaturated
bonds are called polyunsaturated fatty acids (PUFA).

As we will discuss in subsequent chapters, there is good reason to believe that both the quantity and the specific nature of the saturated and unsaturated fats in the diet are important factors in a number of different diseases. For example, it is for this reason that the American Heart Association has for many years advocated a reduction in the percentage of the total calories derived from fat, together with a decrease in the ratio of saturated to unsaturated fat in the diet.

Finally, it should be understood that a given fat is a very complex mixture; butterfat alone contains more than 500 different fatty acids. For this reason, one must usually speak in rather general terms about the various kinds of fats and oils that we eat. Several statements may be made:

(1) Animal fats tend to be saturated fats.
(2) Vegetable fats tend to be unsaturated.
(3) The greater the degree of saturation, the more likely it is that the fat will be a solid at room temperature.

There are some exceptions to the general rule that vegetable fats are relatively unsaturated. For example, the respective degrees of saturation of the oils of the coconut, palm kernel, and palm are 90, 85, and 50%. By comparison, butterfat is 62% saturated, whereas the fats of a partially hydrogenated corn oil margarine are still only 16% saturated. The appeal of the saturated vegetable oils for makers of packaged cereals, crackers, and snack foods is that they impart a pleasant taste and they have a longer shelf life. Until recently, not much attention was paid to the fact that some vegetable oils might be just as saturated as butterfat. For example, the *East West Journal*, whose pages advocate "natural" foods, cures, and ways of life, carries ads for "non-dairy

---

...in fact, certain forms of dietary fat are as essential to our health as are proteins or any of the vitamins.

---

*Phil Sokolof...brought national attention
to the issue of saturated vegetable fats....
by the Spring of 1989 many of the largest food
companies in America had...removed tropical oils
from their products.... Whenever an individual can
bring a huge industry to its knees in an apparently
worthy cause, praise is deserved.*

---

coconut milk ice cream" for those who wish to avoid regular ice cream, presumably because of its saturated fat content.

Phil Sokolof is the person who brought national attention to the issue of saturated vegetable fats. In 1988, he was a wealthy, 65-year-old manufacturer of building materials and a freshly recovered heart attack victim. He had already devoted considerable effort and money to developing cholesterol education and testing programs in his home state of Nebraska. When he became aware of the widespread use of coconut and palm oils by national food processors, he decided to do something about it. His campaign culminated in full page ads in some of the country's largest newspapers. The theme of the ads, "The Poisoning of America," was certain to capture the attention of the public. The food processors took note as well; by the Spring of 1989 many of the largest food companies in America had either removed tropical oils from their products or had announced plans to do so.

Whenever an individual can bring a huge industry to its knees in an apparently worthy cause, praise is deserved. However, there are two things that make me think that Mr. Sokolof's triumph may be more symbolic than nutritionally significant, more relevant to honesty in labeling than to any form of food "poisoning." First of all, even if saturated fats were poisonous, which they are not, tropical oils provide only a small part of the total saturated fat in a typical diet; meat and dairy products are far more significant contributors. (A cynic might suggest that it is easier for a son of Nebraska to argue against foreign oils than to condemn beef and full-fat milk.) Secondly, I'm not so sure we know all there is to know about the influence of various fats and oils on heart and vascular disease. For example, in a recent study done in The Netherlands, Margaret Rand and her colleagues found that palm oil differed little from sunflowerseed oil, a highly polyunsaturated fat, in its ability to prevent the formation of blood clots. Nonetheless, Mr. Sokolof has done a great service by moving closer the day when all processed foods will carry complete information about their contents.

A liquid unsaturated fat—an oil—may be converted into a solid saturated fat by the process of hydrogenation in which hydrogen is added to the unsaturated bonds of the fatty acids. If you look at the label of a typical corn oil margarine, you will see that it contains "partially hydrogenated corn oil." This means that corn oil, a predominantly unsaturated vegetable fat, has had some of its unsaturated fatty acids converted to the saturated state. The degree of hydrogenation can be controlled to give a product of just the desired consistency at room temperature. An emerging issue is whether hydrogenation also introduces "unnatural" fatty acids and what the consequences, if any, of such fatty acids might be.

In the previous chapter I described how plant or animal protein is broken down into amino acids, absorbed in the intestine, and reassembled in the body as human protein. A

Sokolof's triumph may be more symbolic than nutritionally significant, more relevant to honesty in labeling than to any form of food "poisoning".... moving [us] closer [to] the day when all processed foods will carry complete information about their contents.

similar process occurs in the digestion of fat. In the digestive tract, fatty acids are broken off the glycerol molecule, absorbed, and then reassembled as human fat.

A curious difference between the handling of fat and protein is that, instead of being acted upon immediately by the liver, most fat is put into suspension as particles called chylomicrons that then enter the bloodstream via the lymph. Chylomicrons give serum, the usually clear component of blood, a milky appearance after a fatty meal. This observation prompted a number of physicians to suggest in the last century that milk might safely be given intravenously as a substitute for blood. In 1878, T. G. Thomas stated in the *New York State Journal of Medicine* that "in injecting milk into the veins we are imitating nature very closely in one of her most simple physiological processes." Thomas predicted "a brilliant and useful future" for the intravenous administration of milk. Despite some glowing tributes to its value, the procedure was soon discarded. We are now aware that the undigested foreign proteins of milk are likely to produce severe allergic reactions.

Are we too fat simply because we eat too much fat? In the middle of the last century many of those few who thought about the question believed that stored fat—adipose tissue—was derived solely from dietary fat. That notion was slowly discarded over a period of about 50 years and it is now known that body fat may be derived from any one of the major components of the diet. Thus, excess intake of protein, carbohydrate, or fat will lead to deposition of body fat. It's for this reason that, in planning a diet for weight reduction, one is concerned not just with fat, but with the entire energy content, usually expressed in calories, of all foods being ingested. We'll discuss these matters in more detail in Chapter 25.

Body fat serves as a buffer against uncertainties in our food supply. When food is plentiful, body fat is laid down against the day when energy intake is inadequate. The role of fat as an energy reserve has been obvious to humankind for many thousands of years. The fact that dietary fat has more subtle and complex functions was not apparent until well into the present century. Recent discoveries with respect to the non-energy reserve functions of fatty acids seem to promise remarkable new treatments for a variety of human diseases.

In 1927, Herbert Evans, whom we shall meet again in Chapter 16 as the discoverer of vitamin E, was aware that rats fed a fat-free diet failed to grow at a normal rate and that vitamin E did not correct the situation. Evans and George Burr, a Minnesotan then working in Evans' California laboratory, suggested the existence of a previously unrecognized vitamin in fat. Their idea of a "vitamin F" was discarded a few years later when Burr and his wife Mildred described a new deficiency disease in rats. They attributed the disease to a lack of a specific polyunsaturated fatty acid, linoleic acid, and coined the term "essential fatty acid." The idea that fat might be a dietary essential gained acceptance only slowly, and it was not until the 1960s that it was fully incorporated into the dogma of human nutrition.

Experiments by Arild Hansen at the University of Texas Medical Branch at Galveston are important for two reasons. First, they provided convincing evidence that certain dietary fatty acids are essential for humans as well as for rats. Second, they illustrated an ethical dilemma often encountered by clinical investigators.

Hansen's work began as a general comparison of the constituents of the blood of well nourished and poorly nourished children. The two groups had about the same total amount of fatty acids, but there was very little linoleic acid in the blood of children who lived in poverty. This suggested to Hansen that linoleic acid might be essential for "the well-being of healthy children." He was of course aware that retrospective studies could not prove the essentiality of linoleic acid—there were far too many other things that might account for the lack of "well being" in poor children. What was needed was a prospective study, one in which groups of infants would be fed varying amounts of linoleic acid and differences in their health could be demonstrated. Just such a study was reported by Hansen and his colleagues in 1963.

A total of 428 newborn children were fed one of four different formulas. Three of these contained 42% of their calories as fat, but their linoleic acid contents were 1.3, 2.8, and 7.3%, respectively. (Human milk provides 4–5% of its calories as linoleic acid.) The fourth formula contained no fat at all. It quickly became evident that babies fed the fourth formula were not growing at a normal rate. Therefore, it was modified to contain 42% of calories as fat, but virtually no linoleic acid.

After three months had passed, the essentiality of linoleic acid had been proven. Babies fed the formulas containing 2.8 and 7.3% linoleic acid appeared to develop normally. In contrast, the infants who received either 1.3% or no linoleic acid at all were retarded in their rate of growth. In addition, the linoleic acid-poor formulas caused the babies' skin to be dry and scaly; in the skin folds there was redness, oozing, and peeling.

As important as were these studies by Hansen and his colleagues in establishing the human need for linoleic acid, today's ethical standards would forbid them. (If 20 years seems too short a time for dramatic changes in medical ethics, consider the alterations in our concept of civil rights over the same time period.) Who were these infant "volunteers?" Most assuredly they were not children of the hospital's professional staff. I assume that all of the babies came from what is euphemistically called a "clinic population," i.e., those receiving "free" medical care; the authors do tell us that 70% were "Negro infants." None of the children could benefit from the study and some were almost certain to be harmed, at least transiently. What then was the motivation of mothers to enroll their children? We are told that "the nature of the study was explained to the mother," but one must won-

> *As important as were these studies by Hansen...in*
> *establishing the human need for linoleic acid,*
> *today's ethical standards would forbid them....*
> *no investigator, however well-intentioned, has the*
> *right to use humans as a group of rats or guinea*
> *pigs would be treated. (Some advocates of animal*
> *rights would turn that around—don't use rats or*
> *guinea pigs as you would a group of humans.)*

der about the extent of that explanation. The answer most likely lies in the fact that "milk and solid foods were provided gratis for the duration of participation in the study." Today we might call that economic coercion.

It is not my purpose here to condemn the late Dr. Hansen; his work was ethical in the context of his times and it surely was an important contribution to the science of nutrition. I wish merely to point out that progress in medical science is often slowed by the still emerging concept that no investigator, however well-intentioned, has the right to use humans as a group of rats or guinea pigs would be treated. (Some advocates of animal rights would turn that around: don't use rats or guinea pigs as you would a group of humans.)

Standard textbooks of nutrition or biochemistry usually list three essential fatty acids (EFAs: linoleic acid, arachidonic acid, and linolenic acid). In fact, the human body can convert linoleic acid to arachidonic acid, so the latter is not a true dietary essential. Because of their widespread distribution in animal and plant tissues, EFA deficiency has never been observed in a normal human being. However, there are many people who are unable either to take food by mouth or to absorb nutrients because of accident, surgery, or disease. What has kept them alive for these extended periods is a technique called total parenteral nutrition (TPN). During

TPN, all nutritional needs are provided in a liquid that is introduced into a large vein.

For many years, no satisfactory source of fat was available for use in TPN and none was thought to be needed because weight gain was often quite satisfactory without it. However, it became evident that skin rash, much like that observed by Hansen in babies, often occurred after three months or so on TPN, and that the rash disappeared if linoleic acid was provided.

Because of their names, it is an easy matter to confuse linoleic and linolenic acids. But our bodies can tell them apart and one cannot be substituted for the other. The effects of a deficiency of linolenic acid were first described in 1982 by Ralph T. Holman and his associates at the Hormel Institute of the University of Minnesota. Dr. Holman's patient was a six-year-old girl who had had a nearly nine-foot length of her small intestine surgically removed following a gunshot wound. Five months after she was started on TPN, she experienced periods of numbness, weakness, inability to walk, pain in her legs, and blurring of vision. It was then discovered that the fat that was being given her contained very little linolenic acid. When more linolenic acid was given, all of the girl's symptoms disappeared.

If essential fatty acids were a significant concern only in patients maintained by total parenteral nutrition, they would remain of little interest to most of us. However, as I mentioned earlier in this chapter, we now know that EFAs are converted in our bodies into chemicals whose functions are only dimly perceived, but whose importance is without question. There is today a real hope and expectation that selective manipulation of essential fatty acids may be useful in the treatment of heart disease, inflammation, asthma, cancer, and many other diseases.

Our story begins in 1930 at Columbia University in the laboratory of Raphael Kurzrok and Charles Lieb. They observed that fresh semen affected the muscle tone of the human uterus. A few years later, Ulf Svant von Euler, a Swedish physiologist, isolated the responsible substance from the sex glands of male sheep and gave it the name "prostaglandin." More than 20 years then passed before the exact chemical nature of the factor became known, and it was realized that there was not just one, but a whole family of prostaglandins.

Had Von Euler's work been done 50 years later, and had it been supported by the United States government, he might have been a candidate for one of Senator William Proxmire's Golden Fleece Awards, a satiric laurel given by the good Senator to those investigators whose federally financed studies he judged to be wasting the taxpayers' money. Who,

---

There is today a real hope and expectation that selective manipulation of essential fatty acids may be useful in the treatment of heart disease, inflammation, asthma, cancer, and many other diseases.

---

after all, save shepherds perhaps, could have any proper interest in sheep sex?

Von Euler was spared the Golden Fleece, but recognition did come in another form. The 1982 Nobel Prize for Physiology or Medicine was given for studies directly stemming from his work. It was shared by Sune Bergstrom and Bengt Samuelsson, two of von Euler's colleagues at the Karolinska Institute in Stockholm, and by John Vane (now Sir John), a British pharmacologist. (The prize would have gone to von Euler himself, but he had already received the Nobel award 12 years earlier for some of his other discoveries.)

But what has any of this to do with nutrition? The answer to that question was provided in 1964 when David Van Dorp and his colleagues in the Netherlands and, independently, the Swedish workers, proposed that essential fatty acids are the precursor substances from which the body manufactures prostaglandins. The words of Sune Bergstrom, Henry Danielson, and Bengt Samuelson were prophetic indeed: "Although our knowledge of their biological effects is still fragmentary, their wide distribution...[suggests]...that the symptoms of essential fatty acid deficiency at least partly are due to an inadequate biosynthesis of the various members of the prostaglandin hormone system."

Chemical characterization of the prostaglandins was made possible by the development in Sweden of exceedingly sophisticated analytical instruments. In contrast, John Vane and his associates in the Department of Pharmacology of the Royal College of Surgeons of England made their crucial contributions using a bioassay system thought by many to be archaic. Bioassay refers to the use of a living tissue or whole animal to determine the activity of a chemical.

In Vane's laboratory a tissue such as the lung of a guinea pig would be bathed with a solution that then was passed over other tissues known to be sensitive to a variety of biological factors. Late in 1960, it was found that, during an allergic reaction, the guinea pig lung releases a factor that can cause blood vessels to constrict. Vane and Priscilla Piper called it rabbit-aorta-contracting substance (RCS). Subsequently it was shown that arachidonic acid was the source of RCS and that its production was blocked by aspirin. These and other experiments led Vane to propose in 1971 that the pain-relieving, antifever, and antiinflammatory effects of aspirin are the result of the inhibition of synthesis of prostaglandins from arachidonic acid.

Vane's aspirin–prostaglandin hypothesis was soon supported by a large body of evidence. However, there remained a number of puzzling facts. No known prostaglandin had the properties of RCS, nor could any known prostaglandin explain the ability of aspirin to inhibit the clotting of blood. These inconsistencies led Sune Bergstrom to search for and find a previously unrecognized product of arachidonic acid.

He called it thromboxane because of its presence in thrombocytes, an element of the blood involved in the formation of blood clots (thrombi). In retrospect it is not hard to see why thromboxane was difficult to identify: it has a half-life of about 30 seconds. Put another way, 3-1/2 minutes after it is produced from arachidonic acid, 99% of it has been converted to other substances.

Thromboxane appeared to offer a remarkably simple approach to a major set of diseases. Clots that form in the blood vessels of the heart may produce heart attacks; clots that form in the blood vessels of the brain may produce strokes. Might it be that aspirin, by blocking the formation of thromboxane from arachidonic acid, could prevent heart attack and stroke? The idea was so attractive that millions began to take aspirin either on their own or at the suggestion of their physicians. Clinical trials were organized throughout the world.

Because a very large number of subjects would be required to prove that aspirin prevents heart attack and stroke in healthy people, all of the initial clinical trials of aspirin involved persons who had either suffered a heart attack or showed evidence of an impending stroke. The National Heart Lung and Blood Institute in this country began a 17 million dollar effort called the Aspirin Myocardial Infarction Study (AMIS). By August of 1976 they had recruited 4524 people who had survived a heart attack. Half were given one gram per day of aspirin (about three tablets) and the other half received placebo.

But there were other prostaglandins to be discovered. On October 21, 1976, long before the Aspirin Myocardial Infarction Study could be completed, Salvador Moncada and John Vane opened what Vane called a new chapter in the prostaglandin story. They had found a prostaglandin called PGX that inhibited the clotting of blood, an effect exactly opposite to that of thromboxane. Furthermore, aspirin inhibited the synthesis of PGX or, as it later was called, prostacyclin. A complicated story indeed. Would the potentially beneficial effects of aspirin in blocking the formation of thromboxane be more than offset by the potentially hazardous effects of aspirin in diminishing the synthesis of prostacyclin? Would heart attack and stroke victims be helped or hurt by aspirin? Had patients and physicians alike been too quick to jump on the aspirin bandwagon? Questions such as these became known collectively as the aspirin dilemma.

By 1980 it appeared likely that most of those who had gambled on aspirin had won. In 1986 the Food and Drug Administration made it official when they approved a new "indication" for aspirin—reduction of the risk of death and/or heart attack in those who had already had a heart attack or who suffered from unstable angina. But I said that "most of those who had gambled on aspirin had won." That there might be losers as well was shown in several studies by a small but disturbing rise in the incidence of hemorrhagic stroke—in which a vessel in the brain bleeds rather than being blocked by a clot.

The use of aspirin after a heart attack is called secondary prevention and is now accepted therapy throughout the world. But what of primary prevention: the use of aspirin in healthy people to diminish the risk of a first heart attack? As was expected, this turned out to be a much more difficult question to answer, and indeed remains one that, even as I write this, has not yet been answered to the satisfaction of all.

In this country, the Physicians' Health Study asked the related question of whether aspirin would reduce death from cardiovascular disease. The subjects were 22,071 physicians, divided about equally between aspirin and placebo treatment. Begun in 1982, the study was to have run until 1990. However, in a dramatic special report in the January 28, 1988 issue of the *New England Journal of Medicine*, it was announced that the aspirin portion of the study was being terminated because a "beneficial effect (of aspirin) on fatal and nonfatal myocardial infarction had been found." In contrast, no beneficial effect was observed in a study reported two days later in the *British Medical Journal*. Their trial had lasted six years and involved 5139 physicians in England.

The design and conduct of the British and American studies were sufficiently different that a difference in outcome was not terribly surprising. For example, the dose of aspirin used in the US was 325 mg (one standard tablet) every other day, whereas in England 500 mg was taken each day. But let us come straight to the practical question raised by the favorable results of the Physicians' Health Study: Should I and every other person in the country take an aspirin every other day? To answer that question properly, we must look a little more closely at the American findings.

First of all, participants in the Physicians' Health Study were a highly selected and very healthy group: no history of heart attack or stroke, no cancer, no liver or kidney disease, no stomach ulcers or gout, and no adverse reactions to the use of aspirin; virtually none were smokers and no women were included. Second, though it is true that the number of heart attacks (99 vs 171) and deaths from heart attacks (six vs 18) were diminished, the total number of deaths in the two groups was exactly the same (44). Finally, six doctors in the aspirin group died of a stroke, whereas there were only two such deaths in the placebo group. Numbers too small to rule out a chance occurrence, but troubling nonetheless. The decision by you and me to take an aspirin every other day for the rest of our lives must be based upon your and my personal histories, upon our physicians' recommendations, and upon our respective fears of heart attack and stroke.

To complete our catalog of the products of essential fatty acids, we must consider another disease, asthma. During an asthmatic attack, breathing becomes difficult because of a narrowing of the bronchioles, the fine tubes that distribute air throughout the lungs. Cortisone, an antiinflammatory drug familiar to many athletes, relaxes bronchioles and is sometimes useful in severe asthma. The discovery that cortisone prevents the release of arachidonic acid from storage sites in the body provided a reasonable explanation for its ability to reduce inflammation. But what of its anti-asthma effects? If they were caused solely by blockade of prostaglandin production, aspirin should work well against asthma; it does not. In fact, aspirin causes asthma in some people.

Might it be that arachidonic acid is converted into substances other than prostaglandins, chemicals uninfluenced by aspirin? Sune Bergstrom and Pierre Borgeat thought that it might and in 1979 they discovered a whole new family of chemicals, the leukotrienes. Suddenly, an old observation made sense. In 1938, Charles Kellaway, an Australian scientist, identified a factor he called slow reacting substance (SRS). During an asthmatic attack, SRS was released in the lungs and caused the airways to contract. Kellaway proposed that SRS was the agent of asthma. Forty-one years later, SRS was shown to be a mixture of leukotrienes.

Asthma is not a simple disease and it is likely to involve both leukotrienes and prostaglandins as well as other substances. Nonetheless, there is reason for optimism. For example, it has been found that persons sensitive to ragweed respond to the ragweed pollen by liberating leukotrienes. Non-allergic persons neither release leukotrienes nor suffer discomfort when exposed to the pollen. Such findings offer the hope that drugs soon will be developed that will influence leukotriene production in such a way as to be useful in asthma, allergic reactions, rheumatoid arthritis, and related inflammatory conditions.

This chapter was not intended to be a lengthy discussion of prostaglandins. But I would open myself to a charge of male chauvinism if I failed to mention the role of prostaglandins in uterine function.

The uterus is a bell-shaped muscle that receives the fertilized ovum, gives refuge to the embryo and fetus, and, at the proper time, delivers a baby. Uterine contractions are normally under the control of prostaglandins. During pregnancy, administration of excess prostaglandins will cause premature contraction of the uterus and the fetus will be expelled. Thus, prostaglandins may be used as agents of abortion. In nonpregnant women there is a monthly cycling of prostaglandin levels. In one-quarter to one-half of all menstruating women, peak levels of prostaglandins cause painful cramps (primary dysmenorrhea). Beginning in the mid-1970s, clinical trials showed that inhibitors of prostaglandin production were very effective in controlling menstrual cramps. The trade names Motrin (ibuprofen), Ponstel (mefanamic acid), and Naprosyn (naproxen) soon became known to millions of women. Recently, the Food and Drug Administration made ibuprofen available without prescription under the names Advil and Nuprin. (Perhaps to remind us of the complexities of the prostaglandin systems, aspirin was found to be relatively ineffective against menstrual cramps.)

Let's now return to nutrition. I mentioned earlier that the concept of essential fatty acids gained general acceptance only quite recently. We shouldn't be surprised then to find no consensus about the optimal amount to be eaten. Intake of essential fatty acids to avoid all signs of deficiency is easier to determine. Patients maintained on total parenteral nutrition get along quite well on 2–7 grams per day. That's not much; two tablespoons of peanut butter or a vegetable oil margarine contain about 5 grams of polyunsaturated fat.

No controversy surrounds the ease or desirability of avoiding deficiency of essential fatty acids. When we turn to questions of the optimal amount of fat in the diet or the best ratio of unsaturated to saturated fats, matters are far less certain. The Tenth Edition of the Food and Nutrition Board's Recommended Dietary Allowances suggests that fat should represent no more than 30% of the total caloric intake. Many believe that percentage to be too high and some think it much too high. A few have even implied that a 30%-fat diet is in the best interests only of those who raise pigs and cattle for a living.

Matters of agricultural politics and flavorful cooking aside, there's no reason to believe that we need any fat at all beyond the essential fatty acids. On the other hand, I don't think that anyone is ready to suggest a diet containing five grams of PUFA and no other fat. Nonetheless, the trend is ever downward. The American Heart Association also believes that 30%-fat is appropriate for healthy Americans. When there is evidence of heart or circulatory disease, the AHA recommendation for total fat drops all the way down to 20%. Indeed, the late Nathan Pritikin seemed to do no immediate harm with a diet that contained only 5–10% fat. We shall consider in detail the proposed relationships between fat, cholesterol, and heart disease in Chapters 22 and 23.

We have come a long way from von Euler's sheep prostates and the Burrs' notion of EFAs to our knowledge of prostaglandins, thromboxanes, prostacyclins, and leukotrienes. It has taken 50 years and what has already been learned permits nearly unlimited speculation. Medical scientists will devote many more years to teasing out that which is true from that which is merely plausible. Fools and charlatans will rush to tell us how our lives can be extended or improved simply by eating more of a particular fatty acid.

Advertising copy for health food stores and vitamin catalogs generally avoids the word fat, except to condemn it. Unsaturated fatty acids are usually referred to as "unsaturates." Fat is mentioned only as something to be lost. However, those who compose these promotional blends of science and fiction are not totally unaware of EFAs and prostaglandins and such. Thus we find more and more frequent use of words such as lecithin, eicosapentanoic, and gamma-linolenic. The story of the evening primrose is worth telling to illustrate the often tenuous connection between science and advertising.

Rats, when deprived of linoleic acid, develop dry and scaly skin. George and Mildred Burr told us that in 1929 when they first expressed the concept of essential fatty acids. Human beings, for reasons largely unknown, often suffer from eczema, a noncontagious, chronic inflammation of the skin. Might it be that eczema is a sign of fatty acid deficiency? Arild Hansen thought it might and, beginning in 1931, he treated his young eczematous patients with corn oil and other unsaturated fats. The results were mixed; some thought it did some good, others did not, and that's where the matter stayed for many years. Then, in 1954, a report appeared in the *British Medical Journal* that seemed to put the matter to rest. John Pettit had designed and carried out a very careful study in children. He concluded that linoleic and linolenic acids, taken by mouth or directly applied to the skin, have no beneficial effect on eczema either alone or in combination with standard treatment. Enter the evening primrose.

Linoleic acid in the diet is converted into arachidonic acid; that's why we don't have to eat the latter. One of the intermediates in the conversion is gamma-linolenic acid (GLA). In 1980, J. L. Burton and his colleagues at the Department of Dermatology of the Bristol (England) Royal Infirmary reasoned that patients with eczema don't respond to linoleic acid because they can't convert linoleic acid to GLA. Sources of preformed GLA are limited; the seed oil of the evening primrose is one. In 1981 and 1982, Dr. Burton and his associates reported their results with a total of 131 children and adults treated with primrose oil. They found a "modest but significant improvement" in eczema when the oil was added to the usual regimen of steroids, antihistamines, and emollients.

Despite the limited beneficial effect seen by Dr. Burton, several facts combined in the 1980s to make evening primrose oil the darling of the vitamin and health food entrepreneurs. First, eczema is a very common condition; it accounts for about one-third of all patients who visit dermatologists. In addition, eczema is seldom cured by presently available methods. Second, the remarkable discoveries by Bergstrom, Samuelsson, Vane, and others make plausible all sorts of therapeutic possibilities for essential fatty acids. Third, the

---

*Mark Twain's comment...seems applicable as well
to some of the commercial purveyors...
of nutritional information: "There is something
fascinating about (it). One gets such wholesale
returns of conjecture out of such a trifling
investment of fact."*

---

use of evening primrose oil has been vigorously promoted
by an organization located in Nova Scotia called the Efamol
Research Institute. (Efamol is the name of a product that
combines primrose oil and vitamin E.) During the last decade, members of the Institute have suggested that Efamol may
have value in preventing or treating conditions ranging from
eczema to cancer, from arthritis to high blood pressure; in
1986, Drs. Begin and Das of the Institute wrote that gamma-
linolenic acid might cure AIDS.

Stimulated by the Efamol Research Institute, the banner
of the evening primrose was taken up by the thousand or so
General Nutrition Centers in the United States. GNC implied
that the oil is useful in "weight control, arthritis, hyperactivity, alcoholism, and much more." GNC promised to keep us
informed "as more exciting discoveries are made." (With
enthusiasm like this, it's easy to understand why David
Shakarian, GNC's founder, was reputed to be Pittsburgh's
richest man when he died in 1984.)

GNC was not alone in their promotional efforts. The
local representative of the Feel Rite Health Food Shoppes,
Inc. gave me an article by one Alan Donald who called primrose oil "a star still rising." Mr. Donald writes that it is of
value not only for all the things GNC mentioned, but also
for heart and circulatory disease, disorders of the immune
system, multiple sclerosis, childbirth (a disease?), premenstrual syndrome, cancer, and brittle nails. Mark Twain's

---

*For those not inclined to dietary gymnastics, a
simple health-respecting measure is to decrease
the intake of beef, pork, and processed meats,
with a compensatory increase in the consumption
of fish and vegetables.*

---

comment about science seems applicable as well to some of
the commercial purveyors (perverters?) of nutritional information: "There is something fascinating about (it). One gets
such wholesale returns of conjecture out of such a trifling
investment of fact."

Summing Up

Although most of us, most of the time, think of fat only
as something "to be lost," certain components of dietary fat,
the essential fatty acids, are actually required for our well
being. Fortunately, deficiency of essential fatty acids is a rare
condition, irrelevant to nearly all of us. Instead, our concern
should be focused on the consumption of too much fat in
general, and too much *saturated* fat in particular. An abundance of evidence indicates that saturated fat is the major
contributor to elevated levels of cholesterol in the blood. Less
convincing evidence links dietary fat to some forms of cancer.

A useful generalization is that animal fat tends to be saturated, whereas the fats in vegetables, usually called oils,
are unsaturated.

For those not inclined to dietary gymnastics, a simple
health-respecting measure is to decrease the intake of beef,
pork, and processed meats, with a compensatory increase in
the consumption of fish and vegetables.

CHAPTER 3

Carbohydrates

From Glucose to Oat Bran to Sawdust

It presently is fashionable to suggest that we increase our intake of complex carbohydrates.

Using the table below as a guide, let us begin with a few definitions and a brief consideration of just how carbohydrates are handled by our digestive systems:

Carbohydrate	Other name	Complexity	Simplest unit
Glucose	Dextrose	Monosaccharide	Glucose
Fructose	Fruit sugar	Monosaccharide	Fructose
Sucrose	Sugar	Disaccharide	Glucose, fructose
Lactose	Milk sugar	Disaccharide	Glucose, galactose
Starch	Plant starch	Polysaccharide	Glucose
Glycogen	Animal starch	Polysaccharide	Glucose
Cellulose	Fiber	Polysaccharide	Glucose

Carbohydrates take their name from the fact that they are made up of carbon and water (hydros). They may be "simple" or "complex," depending upon how many individual units are strung together. Glucose and fructose are simple carbohydrates; were they any simpler, they would lose their outstanding sensual characteristic, a sweet taste. When glucose and fructose are joined together, a new carbohydrate, sucrose, is formed. Although sucrose is commonly called "sugar," confusion will be avoided if we remember that all of the many simple carbohydrates with a sweet taste are sugars.

37

---

*...the "complex carbohydrates"...are indigestible; our
bodies cannot liberate their sugars; they add little to our
caloric intake. Such carbohydrates are also called fiber,
a buzz word of pop nutrition of the 80s.
Strange advice indeed: eat foods that you can't digest.
Stranger yet, it's good advice.*

The Latin word for sugar is saccharum, so we sometimes refer to sugars as saccharides; glucose and fructose are thus monosaccharides, sucrose a disaccharide, and so forth. When many units of a simple carbohydrate such as glucose are strung together, we speak of a polysaccharide or complex carbohydrate.

It presently is fashionable to suggest that we increase our intake of complex carbohydrates. This may be a bit unsettling for those of my generation, who as children in the 1940s and 50s were told to cut down on "starchy foods"; starch is a complex carbohydrate. To make sense of this, we need be aware that though all complex carbohydrates are made up of sugars, our bodies are not able to digest them all equally well. We are extremely efficient in liberating glucose from starch, though a polysaccharide such as cellulose is quite indigestible, unless you happen to be a cow.

We have seen previously how dietary protein and fat are broken down in the gut, absorbed, and then reassembled in the body as human protein or fat. The same thing happens with digestable carbohydrates. Starch is hydrolyzed to glucose and absorbed. The glucose is then either used directly as an energy source, or reassembled in the liver into another complex carbohydrate called glycogen. Glycogen is the human equivalent of plant starch. Glycogen and starch do not differ in their basic unit, glucose, but only in the arrangement of those units.

Just as fat is our major long-term energy reserve, glycogen is our short-term stockpile. Very few of us have a constant intake of food. Instead the body is continuously adjusting to meals and between meals. Glycogen is assembled or disassembled in concert with an excess or deficiency of glucose. In a process called glycogen loading, marathon runners and other endurance athletes go through a pre-race ritual in which they attempt to maximize their stores of glycogen.

But what of the current advice to increase consumption of complex carbohydrates? If I simply eat more starch than my body needs to fill the glycogen stores, the excess glucose will be converted to fat. This is not what is intended. Instead, the "complex carbohydrates" of current dietary advice are indigestible; our bodies cannot liberate their sugars; they add little to our caloric intake. Such carbohydrates are also called fiber, a buzz word of pop nutrition of the 80s. Strange advice indeed: eat foods that you can't digest. Stranger yet, it's good advice. We shall see why in a moment.

The human body depends upon glucose for energy. Carbohydrates are the only dietary source of glucose. Yet, there is no Recommended Dietary Allowance for carbohydrates. The reason for this is that our bodies can convert protein or fat into glucose. However, the use of protein and fat for this purpose is generally undesirable. Protein, especially animal protein, is too expensive and fat, especially saturated fat, is a likely contributor to heart disease and other maladies. Although there is no general agreement on the ideal balance between dietary fat, protein, and carbohydrate, many now believe that the safest source of energy is carbo-

---

Although there is no general agreement
on the ideal balance between dietary fat, protein,
and carbohydrate, many now believe that
the safest source of energy is carbohydrates.

---

*Scientists are never comfortable in saying
that this or that statement is utter nonsense....
However, the notion that dietary sugars,
particularly sucrose, are poisonous substances
certainly approaches utter nonsense.*

---

hydrates. Surveys have indicated that the typical American consumes 46% of his or her energy as carbohydrates. The diet the American Heart Association suggests for those with high cholesterol recommends 50–55%, and dietary extremists such as the late Nathan Pritikin advise 70–75%.

Scientists are never comfortable in saying that this or that statement is utter nonsense. So unwilling to take firm positions do they seem that philosophers have advised them not to leave their minds so open that their "brains fall out." However, the notion that dietary sugars, particularly sucrose, are poisonous substances certainly approaches utter nonsense.

As we've already seen, sucrose is a simple carbohydrate, a disaccharide. Sucrose entering the small intestine is rapidly broken into its component parts, glucose and fructose, and it is these monosaccharides that are absorbed. Sucrose is found in a variety of fruits and vegetables. When sucrose is separated from the fibrous parts of a plant such as the sugar cane, it is called refined sugar. This pure white crystalline material is, aside from glucose itself, as direct and uncomplicated an energy source for the human body as one can imagine. In the next chapter I'll provide some detail on the reasons that sugar and poison have became connected in the minds of many. An equally interesting question, and perhaps one of greater nutritional significance, is how the indigestible remnants of plant cells—the fiber—have come to be regarded as a panacea for all human woes.

Sometime in prehistory, humans began to eat cooked cereals as porridges. They also found that grinding cereal

---

*For the Greeks and Romans... dark fiber-laden
bread was for slaves; bread made from fine white
flour was fit for the gods or a king....Hippocrates,
the most famous of ancient physicians, knew that
dark bread "clears out the gut"...[and] conclude[d]
that dark bread was less nutritious than white.*

---

grain between rocks allowed the separation of digestible from indigestible parts of the grains. For the Greeks and Romans of 2000 years ago, dark fiber-laden bread was for slaves; bread made from fine white flour was fit for the gods or a king. And Hippocrates, the most famous of ancient physicians, knew that dark bread "clears out the gut." The added bulk of the stool led him to conclude that dark bread was less nutritious than white. This view was preserved by Galen, a Greek physician of the 2nd century AD, and both Hippocrates and Galen were cited by physicians well into the 20th century.

Not everyone agreed with Hippocrates. The young Plato thought whole meal bread, the bread of the working class, to be superior to the fine white bread of the rich. Socrates regarded whole meal bread as fit only for pigs. An apocryphal story has it that Boerhaave, a celebrated physician of the 18th century, left behind a book that contained all the secrets of his medicine. Every page was blank but one. On it was written: "Keep the head cool, the feet warm, the bowels open."

---

*Not everyone agreed with Hippocrates.
The young Plato thought whole meal bread,
the bread of the working class,
to be superior to the fine white bread of the rich.
Socrates regarded whole meal bread as fit only for pigs.*

Each generation has fought the battle anew. Without the intervention of science, the argument could never be settled. The proponents of a "return to nature" sought your faith; they could provide no evidence. Sylvester Graham is an interesting representative of the evangelistic school of nutrition. Born in Connecticut in 1794, he studied for the ministry before shifting his attention to the hygiene of the body rather than of the soul. He advocated unsifted whole-wheat flour and gave his name to the Graham cracker. Among his prescriptions for a long and healthy life were loose clothing, hard mattresses, cold showers, and no alcohol.

During World War II, T. L. Cleave was the ship's doctor aboard HMS King George V. One of the most common medical problems he encountered among the sailors was constipation. Just as had the ancient Greeks before him, Cleave treated the condition with crude bran, a form of fiber that is almost entirely composed of cellulose. From this humble beginning, Cleave developed the idea that many human diseases are the consequence of improper diet. More specifically, he proposed that when carbohydrate crops—whether grains or fruits or vegetables or sugar cane—are consumed in large quantities, they produce disease. His list of diseases included varicose veins, deep venous thrombosis, hemorrhoids, dental decay, obesity, diabetes, coronary heart disease, appendicitis, diverticulitis, gall bladder disease, stomach ulcers, gout, hypertension, hiatal hernia, and acne rosacea; there is something here for nearly everyone.

In the years after the war, Dr. Cleave published a series of short books in which he developed his views. In the early 1970s, his final book appeared. It was titled simply *The Saccharine Disease*. The medical establishment paid little attention to Dr. Cleave and *The Saccharine Disease*. He was neither a proven scientist nor a distinguished physician. If there existed a rational basis for his views, he did little to make it known. Indeed, his idea that human food should be as close to its natural state as possible was unchanged from that held

---

Boerhaave, a celebrated physician of the 18th century,
left behind a book that contained
all the secrets of his medicine.
Every page was blank but one. On it was written:
"Keep the head cool, the feet warm, the bowels open."

---

by a minority of the population for thousands of years. Cleave's uncritical attraction to what he thought "natural" often led him astray. For example, he believed that animal fat in the diet could do no harm since it was both natural and unrefined; all now reject that belief. Cleave argued that undigested material in the large intestine leads to disease; as we shall see in a moment, many benefits are now attributed to the indigestible parts of the diet.

Concurrent with Dr. Cleave's development of the hypothesis of a "saccharine disease," a group of his fellow British-trained physicians was gathering evidence that would lead to quite different conclusions. More important, their ideas would be solidly rooted in observation, unencumbered by any general philosophy of life or of medicine.

During World War II, Dr. Alexander Walker of the South African Institute for Medical Research in Johannesburg began to observe the effects of high fiber diets on stool bulk and frequency of defecation in the Bantu of South Africa. In addition, Walker observed that despite a number of dietary deficiencies, the Bantu rarely suffered a large number of diseases that caused much sickness and death among Europeans. For example, heart disease was much less common among the Bantu. Hugh Trowell of Uganda visited Walker in 1958 and became convinced that fiber influenced diseases of the colon. At about the same time, Dr. Neil Painter reached much the same conclusion and suggested that diverticular disease of the colon be treated with a high fiber diet. (It is remarkable how often diseases are treated by means later

---

Dr. Denis Burkitt's insight and advocacy...
moved "the fiber hypothesis" from obscurity to the
forefront of scientific and popular nutrition....
Fiber was the answer. The "diseases of Western
civilization" weren't caused by a mysterious
poisonous action of sugar. They were
diseases of deficiency—a deficiency of fiber.

---

shown to worsen the condition. More remarkable still is the longevity of such treatments; a low-fiber diet was standard treatment for diverticular disease for 50 years prior to Painter's work.)

In 1960 Trowell published a book listing eight colonic disorders and 32 other "non-infectious diseases of unknown origin rarely reported in rural blacks but common in Western populations." Trowell, Painter, and others began to use the collective term, "diseases of civilization." The good scientists among them were quick to point out that fiber intake was hardly the only difference between natives of rural South Africa and participants in "Western civilization." Many of our present apostles of fiber have forgotten (or never learned) that simple fact.

Denis Burkitt assured himself a knighthood and a place in medical history when in 1957 he described a peculiar malignant disease of the lymphatic system in Ugandan children that is now known as Burkitt's lymphoma. Using a network of 140 hospitals in Africa and Asia, he set out to study the epidemiology of the disease. As the years passed and the data were collected, Burkitt found himself more and more intrigued by odd distributions of diseases of the colon and of the venous system.

In 1967, Sir Richard Doll, Professor of Medicine at Oxford, introduced Dr. Burkitt to Dr. Cleave and his notion of a saccharine disease. Suddenly, the indecipherable complexity of the African epidemiology of colonic and venous disorders was made simple for Burkitt. Fiber was the answer. The "diseases of Western civilization" weren't caused by a mysterious poisonous action of sugar. They were diseases of deficiency—a deficiency of fiber.

It was Burkitt's insight and advocacy that in large measure moved "the fiber hypothesis" from obscurity to the forefront of scientific and popular nutrition. For American medicine, the turning point was the appearance of a paper by Burkitt, Walker, and Painter in the *Journal of the American Medical Association* for August 19, 1974. It was titled simply "Dietary Fiber and Disease." In the decade following its publication, a consensus developed regarding fiber. Simply stated, the American population would benefit from an increased intake of fiber. That consensus often has been misinterpreted by popular writers and by those in the "health food" movement. First of all, until the mid-1980s, fiber was usually equated with cereal bran. Secondly, it has been assumed that the addition of bran to a Western diet will convert an American, dietarily speaking, into a rural African.

Fiber, however, is not a uniform material. It is composed of many components, the specific effects of which are still largely unknown. It is useful to divide plant fiber into two categories based upon water solubility. The prototypical insoluble fibers are wheat bran, a carbohydrate, and lignin, a noncarbohydrate. The water-soluble carbohydrate fibers include pectin and a variety of plant materials called gums. Though it is true that citrus fruits are rich in pectin and that oat bran is an abundant source of gums, we should not forget that all fruits and vegetables present a complex mixture of fibers. For example, the much maligned potato is a good source of both lignin and pectin.

Division of fibers into soluble and insoluble categories is of considerable practical significance. This is illustrated by an investigation recently conducted by Drs. K. L. Wrick and Daphne Roe and their colleagues at Cornell's Division of Nutritional Sciences, who studied the effects of cabbage fiber (a rich source of pectin), purified cellulose, and wheat bran, ground either coarse or fine, in healthy men over a period of 80 days. The four sources of fiber had quite different effects on mouth-to-anus transit time, stool bulk, and ease of defecation. Fine-ground bran and purified cellulose were associated with hard, dry, difficult-to-pass stools. In contrast, coarse bran and cabbage fiber made for ease of defecation despite the fact that they had quite different effects on stool bulk. To understand these results, we need briefly to consider the mechanics of stool formation and some of the strange forms of life that inhabit our bodies.

In the mid-1980s there arose a large and very successful company called Herbalife. Their products were "all natural," i.e., food-derived, and hence virtually immune from government regulation. One of Herbalife's products was intended to "cleanse the digestive tract." Herbalife was not alone; the notion that the human digestive tract is in constant need of cleaning pervades the natural foods movement. For example, Leonard Jacobs addressed himself to the issue of "colon health" in the September 1984 issue of *East West Journal*. We are told that uncleaned colons contribute to "constipation, diarrhea, tendency toward getting colds, allergies, intestinal gas, lower back pain, swollen belly, body odor due to putrefaction in the intestines, tendency toward overeating while not being satisfied, inflexibility both physically and mentally, lack of stamina, impatience and short temper." It seems that we all need the human equivalent of a Roto Rooter. Alas, the notion that a "clean" digestive tract is essential, as intuitively attractive as the idea may be for most of us, is in fact unnatural, unhealthy nonsense.

Upon first feeding by breast or bottle, humans begin to develop a population of bacteria that soon exceeds in number the cells of the body itself. The quantity and kinds of these bacteria rise and fall with the nature of the diet. A single ounce of the colonic contents may contain 300 billion bacteria representing any of some 400 species. Their specific contributions to human life remain largely unknown. One thing is certain, they are largely anaerobic; they derive their energy for growth and multiplication by fermentation, the same process that causes bread to rise and beer to brew and corn mash to become whisky.

Now let us return to coarse bran and cabbage fiber. These illustrate the two general methods for prevention of constipation. Coarse bran is left largely undigested and un-fermented, absorbs water, and results in a bulky, soft stool. Transit time is decreased and it is likely that daily evacua-tion will result. Cabbage fiber produces a soft stool, but with considerably less bulk and with little effect on transit time. The reason for this difference is that cabbage fiber is exten-sively fermented by the bacteria of the gut. The bacteria multiply enormously and the stool that is passed is in large measure made up of the soft bodies of these bacteria with their high water content.

If coarse bran and cabbage fiber are so different in the way they are handled by the gut, a question that naturally comes to mind is "which form of fiber is preferable?" Should we add generous amounts of crude fiber to our usual diet, or should we obtain our fiber from a variety of cereals, fruits, and vegetables? Should we emphasize water-soluble or water-insoluble fiber? If ease of defecation were the only point, it would make little difference. Defecation is not the only point. For example, heart disease is much less common among the Bantu than in Western societies. In Chapters 22 and 23 we will examine the role played by soluble fiber in the maintenance of appropriate levels of cholesterol in the blood. In September 1984 the National Cancer Institute be-gan a massive educational effort aimed at preventing the occurrence of cancer. They suggested that typical Ameri-cans double their daily intake of fiber from 10–20 to 25–35 grams. This increase was to take the form of increased consumption of the foods listed below. You will note that crude fiber is not included as such. Instead all forms of fiber are ingested as parts of their natural sources. The fact is that we do not know how fiber intake acts in most or all cases. We do know that no epidemiological study has ever found an association of the intake of isolated crude fiber with diminished incidence of a disease. The association has always been with foods that contain fiber.

Foods Rich in Fiber

Whole grain products:

(a) Crackers, bran muffins; brown, rye, oatmeal, pumpernickel, bran, and corn breads; whole-wheat English muffins and bagels.
(b) Breakfast cereals such as bran cereals, shredded wheat, whole grain or whole-wheat flaked cereals, and others that list dietary fiber.
(c) Other foods made from whole-wheat flours, including barley, buckwheat grouts, bulgar wheat, macaroni, pancakes, pasta, and taco shells.

Fruits and vegetables:

(a) Apples, pears, apricots, bananas, berries, cantaloupe, grapefruit, oranges, pineapples, papayas, prunes, raisins, and others.
(b) Artichokes, carrots, broccoli, potatoes, corn, cauliflower, Brussels sprouts, cabbage, celery, green beans, parsnips, kale, spinach, other greens, yams, sweet potatoes, turnips.

Dried peas and beans:

(a) Black, kidney, garbanzo, pinto, navy, white, and lima beans.
(b) Lentils, split peas, and black-eyed peas.

Variety, variety, variety; that's what's needed. The relationship of dietary fiber to cancer and heart disease nicely illustrates the virtues of variety. Epidemiologic evidence suggests that insoluble fiber, the kind found in wheat bran, whole-wheat products, brown rice, and asparagus, is protective against cancer of the colon. On the other hand, it is likely that any cholesterol-lowering effect of fiber is caused by soluble forms of fiber found in foods such as oat and rice bran, carrots, berries, corn, peas, and so on. Until we know in intimate detail the interactions of diet with health and disease, we are foolish to try to put together an optimal diet by loading up on a wide variety of "supplements," including supplemental fiber, whether soluble or insoluble. Take your fiber in all kinds of cereals and fruits and vegetables that appeal to you and get all of the other goodies, known and yet to be discovered, that come with it.

The virtues of a high-fiber diet now seem obvious to most. However, a couple of minor reservations need to be expressed. Fermentation produces carbon dioxide and smaller amounts of other gases. In general, the quantity of gas formed is a function of the nature of the fiber and of the colonic bacterial population resident at the time of consumption. However, the system as a whole is in a constant state of flux. Excessive gas formation seems to reflect changes in diet as much as any specific components. Thus, travelers adopting a new diet are often in distress until the bacterial order of their colons reaches a new equilibrium. (The distress is almost universally attributed to "food poisoning.") In any event, most of us develop a pretty good idea of what is tolerable to one's digestion and what is not. So long as we introduce new sources of fiber in moderation, the problem of "gas" should be minor.

A second concern regarding a high-fiber diet is that absorption of other nutrients may be decreased. This is not a new idea. The ancient physicians were nearly unanimous in believing that dark bread was less nutritious than white because so much more of it appeared as waste. When science was first applied to the question in the early 1800s, the larger amounts of protein, fat, and minerals found in whole wheat breads seemed to be balanced by increased fecal loss; white versus dark looked like a draw.

Oddly enough, our present understanding of the question of fiber and nutrient absorption originated not in direct studies, but in the observation more than 30 years ago that chickens fed soybean protein did not thrive. Despite the fact that soybeans contain adequate amounts of zinc, only the addition of zinc to the chickens' diet returned their growth rate to normal. The answer to this puzzle was provided in 1960 by B. L. O'Dell and J. E. Savage of the Departments of Agricultural Chemistry and Poultry Husbandry of the University of Missouri. They demonstrated that a component of soybean meal, phytic acid, combined with zinc to make it unavailable for absorption.

Phytic acid is found in many, but not all, of the same foods as is dietary fiber. Examples of high-fiber-phytate foods are barley, dry beans, corn, oats, peanuts, peas, wheat, and potatoes. In contrast, green vegetables are good sources of fiber, but contain little or no phytate. In addition, the level of phytate is diminished by baking and by fermentation. The latter may be in the form of leavening of bread doughs or the fermentation of soybean and cereal products such as miso.

The impact of fiber in general, and phytate in particular, on human nutrition is difficult to assess. The most commonly expressed concern is that individuals with low intakes of zinc, iron, or calcium may be thrown into overt deficiency by high levels of phytate. Harold McGee in his book *On Food and Cooking: The Science and Lore of the Kitchen* attributes an outbreak of rickets in Dublin during the Second World War to a combination of low calcium intake and the eating of whole grain breads. Occasionally the natural foods press will exhort us to "fight phytate" or a seller of mineral supplements will vaguely warn of the hazards of phytate. On the

---

Carbohydrates are, or should be, the primary source of calories in the foods we eat....
Although much remains to be learned...
it is difficult to argue against a shift in dietary content away from excess fat and protein to a variety of both simple and complex carbohydrates.

---

other hand, Drs. Ernst Graf and John Eaton have made the interesting suggestion that phytate's ability to bind iron in the colon may be responsible for the lesser incidence of colon cancer in areas where high fiber diets are consumed. In any case, there is no evidence that a varied high-fiber diet need be supplemented or that we need to go to war against phytate.

Summing Up

Carbohydrates are, or should be, the primary source of calories in the foods we eat. But if energy were the only thing we were looking for in a carbohydrate, a simple sugar such as sucrose would do as well as the complex carbohydrates found in cereals, fruits, and vegetables. There are two reasons why, in this case, complex is better than simple. First of all, sucrose is a pure source of energy and brings with it no other nutrients. Second is the concept of dietary fiber, indigestible components of the diet that nonetheless appear to be important in human nutrition. Insoluble fiber, the kind that passes through the body unchanged, may diminish the incidence of diseases of the digestive tract including colon cancer. Soluble fiber, that which is fermented by bacteria that live in our intestines, may be a factor in maintaining a healthy level of cholesterol in the blood. Although much remains to be learned about these matters, it is difficult to argue against a shift in dietary content away from excess fat and protein to a variety of both simple and complex carbohydrates.

CHAPTER 4

Is Sugar a Poison?

The single dietary substance most often condemned by present day evangelists of nutrition is refined sugar. Much of the rhetoric can be traced to Dr. Cleave and his notion, which we considered in Chapter 3, of a "saccharine disease." Perhaps the greatest popularizer and corrupter of the ideas of Dr. Cleave was William Dufty. His book, *Sugar Blues*, was published in 1976 and became a best seller; well over a million copies were printed.

Dufty defines sugar blues as "multiple physical and mental miseries caused by human consumption of refined sucrose—commonly called sugar." Sugar blues includes "depression or melancholy overlayed with fear, discomfort, and anxiety." A few quotations from the paperback cover will give you the flavor of the book. "Like opium, morphine, and heroin, sugar is an addictive destructive drug....If you are overweight or suffer from migraine, hypoglycemia, or acne, the plague of the Sugar Blues has hit you...Exposing sugar, the killer in your diet... " Dufty's book is a sometimes entertaining mixture of science, pseudoscience, hyperbole, name dropping, flimflam, and just plain ignorance; ignorance of physiology, biochemistry, nutrition, pharmacology, and the ways of science in general.

We've already seen in Chapter 3 that sucrose is a naturally occurring substance found in many fruits and vegetables. When concentrated and purified from the sugar beet or sugar cane, it becomes a pure source of energy and for many humans a very attractive substance. When eaten it is

---

*The single dietary substance most often condemned
by present day evangelists of nutrition is refined sugar....
my intention [is] to reject as nonsense the idea...that
sucrose is a poison and a sugar-free diet
the royal road to physical and mental health.
There are quite enough things wrong with the American
way of eating; we need not manufacture problems.*

---

rapidly split into glucose and fructose, absorbed, and either
used immediately for energy or stored for future use. It most
certainly is not my intention to overly defend sugar as a
component of the diet. Sucrose provides energy without
anything else; so-called empty calories; no vitamins or minerals accompany the eating of refined sugar. But it is most
certainly my intention to set out the facts regarding the relationship of sugar to various diseases. And it is most certainly my intention to reject as nonsense the ideas of those like
Dufty that sucrose is a poison and a sugar-free diet the royal
road to physical and mental health. There are quite enough
things wrong with the American way of eating; we need not
manufacture problems.

I want now to consider briefly the following conditions
that have been linked to sugar: obesity, diabetes, heart disease,
dental decay, hypoglycemia, and behavioral disturbances.

Sugar and Obesity

To maintain a constant body weight, our intake of energy in the form of food must exactly equal our expenditure of
energy. Too little intake and parts of the body will begin to
be consumed for energy: first, our stored glycogen is burned,
then our muscle and fat—and we lose weight. On the other
hand, when our energy intake exceeds the body's real de-

mand, the excess will be stored, first as glycogen then as fat—and we gain weight. Too much food, call it excess caloric intake if you wish, produces a gain in body weight. And such gains will occur whether the excess energy is in the form of simple or complex carbohydrates, protein, fat, or alcohol. No convincing evidence has ever been presented that sugar has a specific role in obesity beyond its caloric content. Numerous studies have compared the distribution of calories between fat, protein, and carbohydrate, both simple and complex, in the diets of fat and slim people; no consistent difference has been found. Fat people quite simply take in more energy than they consume; the reasons why they do so are far from simple and will be considered in detail in Chapter 25. Sugar is not the cause of obesity.

Too much food...produces a gain in body weight....
whether the excess energy is in the form of...
carbohydrates, protein, fat, or alcohol.
No convincing evidence has ever been presented that
sugar has a specific role in obesity beyond its caloric content.

Sugar and Diabetes

Except late in the course of starvation, the brain is totally dependent upon glucose as a source of energy. For this reason, our bodies have developed elaborate mechanisms to insure that glucose is available at all times and in the face of fluctuations in dietary intake. Thus energy storage in our bodies is achieved by so-called anabolic processes: the conversion of raw glucose to deposits of glycogen or fat. On the energy supply side, when dietary glucose is inadequate, our catabolic processes come into play: stored glycogen is broken down (hydrolyzed) to glucose, and the amino acids from our proteins provide the energy to convert fat to glucose.

Orchestrating these changes is a set of hormones, of which insulin is the most widely known.

When a meal is eaten, the pancreas secretes insulin, which facilitates anabolism (energy storage): fat, amino acids, and carbohydrates not needed immediately for energy or structural repair are stored against future needs. During a period of fasting, the process is reversed: insulin secretion declines, other hormones are released, and there is a net conversion (catabolism) of stored material to energy.

The diseases we call diabetes involve abnormalities in glucose control. In the total absence of insulin secretion, glucose is unable to enter our cells. As a result, the cells are starved for energy despite the presence of very high levels of glucose in the blood. Eventually that excess is reflected in excretion in the urine (diabetes refers to urine flow and mellitus means honey-sweet). Before the discovery of insulin in 1922 by Frederick Banting and Charles Best of Toronto, people unable to secrete insulin led very short lives. With insulin it became possible to keep patients alive for extended periods. Only then did the full consequences of the disease become apparent. (As the bumper stickers say, "Insulin is not a cure.") The secondary effects include deterioration of function in many areas, including the heart, kidneys, eye, and circulatory system in general. The causes of these secondary effects of diabetes are unknown.

The kind of diabetes that I've just described goes by several names: insulin-dependent, juvenile-onset, type I. By whatever name, it is a disease of unknown origin and, beyond the use of insulin, of uncertain therapy. Nearly 85–90% of those Americans who are called diabetic do not suffer from type I. They are quite able to secrete insulin but, despite that fact, have elevated levels of glucose in the blood. They are said to be insulin-resistant and this form of the disease is called adult-onset, insulin-independent, or type II.

Consumption of carbohydrate causes a rise in the level of glucose in the blood. Diabetes is a disease in which there

---

I wish it were true that excessive consumption of sugar-caused diabetes. Type I diabetics could then easily be spared a lifetime of insulin injections and all diabetics would escape the specter of blindness, kidney failure, and multiple amputations in old age.

---

is an elevated blood glucose. What could be simpler than to conclude that carbohydrates, especially simple sugars like sucrose, cause diabetes? Indeed this notion is so simple and so commonsensical that it is widely believed to be true. (Mother to child: stop putting so much sugar on your cereal or you'll become a diabetic like grandma.)

I wish it were true that excessive consumption of sugar caused diabetes. Type I diabetics could then easily be spared a lifetime of insulin injections and all diabetics would escape the specter of blindness, kidney failure, and multiple amputations in old age. The fact is that type I diabetes is a disease of unknown origin. Hypotheses presently in favor suggest that it results from a viral infection or that it is an autoimmune disease. Likewise, the cause of type II diabetes is unknown, but the outlook for its control is much more favorable. Type II diabetes is closely associated with obesity. Indeed, many now believe that type II diabetes is caused by excess body fat in combination with genetic or disease factors. As a practical matter, obese type II diabetics who lose weight often lose their diabetes as well. A program of aerobic exercise such as the one described in Chapter 20 has at least two potential benefits. First, it can help in the effort to lose weight and, second, aerobic exercise often has a favorable effect upon insulin resistance even in the absence of weight loss.

The history of diabetes and its treatment provides several examples of what Mark Twain called "things we know that ain't so." I've already mentioned the attractiveness of the idea that diabetes is caused by eating carbohydrates, especially sugar. It should not surprise us that this notion has played a prominent and almost equally misguided role in the treatment of the disease as well. A British physician, John Rollo, is credited with launching the era of carbohydrate restriction in diabetes in 1798; that era has not yet passed completely. In 1916, Elliot Joslin wrote that "one likes to think that diabetes is due to the pancreas being overworked and tired out." To correct that situation he recommended a low carbohydrate diet. (Joslin, then an assistant professor of medicine at the Harvard Medical School, later gave his name to Boston's Joslin Diabetes Clinic.)

If we are to reduce carbohydrate intake in diabetes, what shall be substituted? The answer in 1916 and for a long time after was fat. The diet of Leonard Thompson is illustrative. On January 11, 1922, Thompson became the first patient to receive insulin. In addition, he remained on the diabetic diet of the day: 70% fat and only 20% carbohydrate. When Thompson died at age 27 of pneumonia brought on by his diabetes, autopsy revealed widespread atheroslerosis and coronary artery disease. His high-fat diet, intended to spare his "tired out pancreas," almost certainly contributed to the sad state of his arterial system.

The dogma that diabetics could not handle carbohydrates made so much "sense" that it was generally accepted until just a few years ago. Some began to question the dogma as early as 1932 and numerous high-carbohydrate diets were subsequently tested and found to be useful and well tolerated. But, it was not until the role of saturated fat in heart disease became clear (*see* Chapter 22) that there was significant movement away from high-fat, low-carbohydrate for diabetics.

By 1979, the American Diabetes association had reached the following conclusions:

    (1) The dietary recommendations for diabetic persons are, in most respects, the same as for nondiabetic persons.

    (2) Ordinarily, the nutrient needs of diabetic persons may

be met without the use of special "diabetic" or "dietetic" foods.

(3) In non-insulin-dependent, obese, diabetic persons, a weight control and exercise program is of primary importance.

(4) Carbohydrate should account for 50–60% of total energy intake.

Within a few years, a similar position was taken by both the Canadian and British diabetes associations. More heresy was to come.

One of the primary goals in the treatment of diabetes is to keep blood glucose levels within normal limits; extreme fluctuations in either direction can produce coma and brain damage. If diabetics are to increase their intake of carbohydrates to 50–60% of total calories, then it makes sense that the carbohydrates be complex; starch rather than simple sugars like sucrose. Evidence that began to accumulate in 1936 and that today is overwhelming indicates that what makes sense regarding blood glucose responses to simple and complex carbohydrates happens not to be true. The reason is that starch is so rapidly converted to glucose in the gut that, for example, eating a white potato has an effect on blood glucose equal to that of an equivalent amount of sugar. This does not mean that a diabetic or anyone else for that matter is well advised to take all their carbohydrates in the form of candy bars and Coke. For the diabetic it does mean that sweetness need not be equated with sin and previously intolerable diets may be made acceptable by including modest amounts of sugar. (Although it is now generally acknowledged that earlier assumptions about blood glucose responses to simple and complex carbohydrates were in error, this area of investigation remains unsettled.) There is in medicine a continuous tension between, on the one hand, the desire to inform patients fully and, on the other, the feeling that many are best served if given only limited details of their disease and its treatment. Dr. Avram Ravina of the Diabetes Clinic of Haifa, Israel provided a nice illustration. "...One diabetic patient asked me whether watermelon was allowed. In view of the (modest) sugar content of this fruit I told him that it might be consumed freely. After a week the patient's wife asked me whether I had allowed her husband to eat four watermelons daily."

[Some believe that] refined sugar is
the major factor in coronary heart disease...
the leading cause of death in...many ...countries....
there is little evidence to support this contention....
Most now believe that consumption
of refined sugar is no more than a marker
for increased affluence and that
the more likely causative factor is saturated fat.

Sugar and Coronary Heart Disease

John Yudkin, Professor of Nutrition and Dietetics at Queen Elizabeth College of the University of London, has contended for many years that refined sugar is the major factor in coronary heart disease (CHD), the leading cause of death in this and many other countries. Unfortunately for Dr. Yudkin and for those who wish to avoid a premature death from CHD, there is little evidence to support this contention. Initially the idea arose from the fact that, in a given country, as the consumption of refined sugar increased, so did the incidence of CHD. Most now believe that consumption of refined sugar is no more than a marker for increased affluence and that the more likely causative factor is saturated fat (see Chapter 22). For example, Venuzuela and Great Britain have about the same per capita sugar intake, but

Englishmen consume much more saturated fat and have a much higher incidence of CHD.

In view of the evidence that most complex carbohydrates are quickly converted to simple sugars and the lack of substantial evidence for a causative link between sucrose and CHD, we might conclude that "high carbohydrate" can be equated with "high sugar." We would be wrong. Once again the reason is not that sugars are poisons, but that refined sugars are too pure; we need all the other things that nature puts with them in their crude state. The sucrose in your sugar bowl is the same as the sucrose in a serving of peas; the fructose in your soft drink is the same as the fructose in an apple; sugar bowls and soft drinks cannot substitute for apples and peas.

One concern that has been expressed regarding the advocacy of high carbohydrate diets is that triglycerides (fat) in the blood will be elevated. Elevated triglycerides have long been viewed as a risk factor for coronary heart disease. There are several lines of evidence suggesting that increased intake of carbohydrates, including refined sugar, does not present a significant hazard in this respect:

(a) Elevated triglycerides are not a potent risk factor.
(b) Elevation of triglycerides after a switch to a high carbohydrate diet may be a transient event.
(c) Any effect of carbohydrates on triglycerides can be mitigated by aerobic exercise.
(d) For a constant caloric intake, a high carbohydrate diet means that the intake of protein and/or fat is diminished. The overall effect is likely to be favorable in terms of blood chemistry.
(e) No epidemiologic study has ever found a relationship between a high carbohydrate diet and increased risk of coronary heart disease. On the contrary, groups that reduce their intake of animal protein and fat and consume high proportions of plant protein and carbohydrate have a decreased likelihood of heart disease.

The precise reason for these epidemiologic observations is unknown, but they lend no support to the idea that a high carbohydrate diet is hazardous to the heart and circulatory system. Indeed, if carbohydrate displaces saturated fat from the diet, the net result will be beneficial.

Sugar and Dental Decay

Tristan da Cunha is by any measure an out-of-the-way place. It lies together with its sister islands of Gough, Inaccessible, and Nightingale in the middle of the South Atlantic Ocean, 2400 miles, give or take a few hundred, from Montevideo to the west and Cape Town to the east; an unlikely spot for a classic study in dental medicine.

Between the World Wars, Tristan da Cunha's 200 or so inhabitants could expect to see a visiting ship about once a year. The diet was ample, but monotonous: fish and a little meat, sea birds and their eggs, potatoes as the major source of carbohydrate, less than two grams of refined sugar per day. The islanders' health was unremarkable, except that their teeth were nearly free of decay. Surveys conducted in 1932 and 1937 by the British Royal Navy found the overall incidence of caries to be less than 10% and the children to be virtually cavity-free.

World War II and the years that followed brought many changes to Tristan da Cunha. The population remained stable, but interactions with the neighboring continents became more frequent. The islanders were no longer dependent upon local foods; they were able to enjoy a diet, and an incidence of dental decay, much like that in England. In 1966, children consumed thirteen times as much refined sugar as before the War and 80% of them had cavities.

Why blame refined sugar? The answer lies in something called the chemico-parasitic theory of tooth decay. In a book published in 1890, W. D. Miller described the decay of teeth that were incubated in a solution of carbohydrate and saliva.

> *...sucrose as a part of a meal is of little importance in causing cavities. An equal amount of sucrose taken in sticky, between-meals snacks is very harmful to the teeth....*
> *Eliminate between-meals consumption of sugars and nearly all dental decay will be eliminated.*

With no carbohydrate, there was no decay. But apply carbohydrate to a sterile tooth and there was again no decay. Miller hypothesized that bacteria in the mouth act upon carbohydrate to produce caries-causing products.

From 1890 to the present, Miller's hypothesis has been tested and refined, but not rejected. We now regard dental caries as a multifactorial disease in which a bacterium, *Streptococcus mutans*, individual characteristics of our teeth and saliva, and diet each plays a role. Eliminate the bacterium (a difficult task, but one that many scientists are now attempting) and there will be no decay. Be born with decay-resistant teeth and a protective saliva (an event clearly beyond our control) and there will be no decay. (In her nearly three decades of life, my elder daughter has not had a cavity despite sharing the diet of her less fortunate parents and siblings.)

Now for something we can do: stop feeding *Streptococcus mutans*. Decay will be dramatically reduced, if not eliminated. It turns out that *S. mutans* is a finicky feeder. Only simple carbohydrates, mono- and disaccharides, are fermented. Furthermore, sucrose is its favorite. Not only does *S. mutans* ferment sucrose to acids that demineralize the teeth, but it also converts some of the sucrose to sticky polysaccharides that prolong bacterial contact with the teeth. This makes a difference. Fructose is as fermentable as sucrose, but only about half as cariogenic because no bacterial glue is produced.

Eliminate sucrose and other simple carbohydrates from the diet and nearly all dental decay will be eliminated. Few of us are willing or able to do that. What of us? A group of mentally retarded adults at the Vipeholm Hospital near Lund, Sweden have provided guidance. Over a period of nearly eight years, Dr. Bengt Gustafssen and his colleagues found that sucrose as a part of a meal is of little importance in causing cavities. An equal amount of sucrose taken in sticky, between-meals snacks is very harmful to the teeth. With this in mind we can modify the advice that opened this paragraph: Eliminate between-meals consumption of sugars and nearly all dental decay will be eliminated. [I am reminded of a 1980s commercial for Snickers, the candy bar. Young people (with beautiful teeth) tell of their between-meals hunger and how a caramel-sticky, 50%-sugar Snickers satisfies them. Dentists are either horrified or delighted by such commercials; it depends upon whether their primary interest is in preventive dentistry or in a busy practice.]

"Natural" foods are not without guilt. Several centuries before the birth of Christ and several millenia before Snickers, Aristotle warned of the dental dangers of soft, sweet figs. Raisins, dates, and other dried fruits are more than half sugar and are sticky besides. In defense of nature, no vegetables exceed 5% sugar and only a few fresh fruits have more than 15%. Ernest Newbrun, Professor of Oral Biology at the University of California at San Francisco, has suggested that no foods containing more than 15–20% sugar should be taken between meals. By that criterion all fresh fruits and vegetables are acceptable and even Coke, Pepsi, and other sweetened drinks, at about 10–12% sugar, are relatively benign. If your children can't survive without cookies or a sugary cereal after school, make sure that the sweet things get washed down with milk.

Sugar, Hypoglycemia, and Behavior

We have already seen something of the mechanisms by which the human body assures that the cells of the brain have an appropriate supply of glucose at all times. In type-I diabetes, glucose rises to dangerous levels (hyperglycemia) because of inadequate release of insulin. But one can have too much of a good thing. Inject a diabetic, or a normal person, with excess insulin and a drastic lowering of blood sugar (hypoglycemia) will result. The possible consequences include loss of consciousness, irreversible brain damage, and death. [The public was recently made aware of the toxicity of insulin by the trials of Claus von Bulow, charged with trying to kill his wife with the substance. Whatever the merits of that charge (juries first found him guilty, then acquitted him on retrial), the unfortunate Ms. von Bulow is in a comatose state from which she is unlikely ever to recover. Once killed by lack of glucose, brain cells cannot be regenerated.]

Each of us has wide swings in blood glucose during the course of a normal day; after a meal the level may be triple that before. Inappropriate labeling of these natural fluctuations as hyperglycemia or hypoglycemia, essentially an invention of diseases, has led to much confusion both within and without the medical profession. In 1974, the august *New England Journal of Medicine* attempted to stem the tide of hypoglycemia by publishing an article by Drs. Joel Yager and Roy Young of the UCLA School of Medicine with the odd title of "Non-Hypoglycemia Is an Epidemic Condition." I doubt that it made much difference. Even fictitious diseases have their uses.

Let's first consider the real disease of hypoglycemia. One possible cause is an insulin-secreting tumor of the pancreas; nature's equivalent of too big an injection of insulin. The consequences can be quite complex and variable, depending on the activity of the tumor and the depths to which blood glucose falls. One set of symptoms is referred to as "neuroglycopenic" (penia is the Greek word for poverty, so neuroglycopenia means poverty of glucose at neurons, a class of cells in the brain). Mild neuroglycopenia may result in fatigue, mental dullness, and headache. When more severe,

> *Fatigue, weakness, confusion, drowsiness, mental*
> *dullness, headache, heart palpitations, tremulousness,*
> *sweating, anxiety: what a marvelous set of*
> *symptoms; so ripe for misattribution....*
> *It is absolutely essential to the diagnosis of*
> *hypoglycemia that symptoms occur when glucose*
> *levels are lowest and that they disappear as glucose rises.*
> *When these criteria are applied, the disease*
> *of hypoglycemia is found to be relatively uncommon.*

there may be abnormal behavior (which has sometimes been confused with mental illness), seizures, and loss of consciousness.

Neuroglycopenic symptoms are a direct consequence of low levels of glucose in the brain. A second set of symptoms caused by hypoglycemia is reflex in nature. As blood glucose falls, the secretion of insulin is turned off, glucose is liberated from storage sites, and the synthesis of glucose in the liver and elsewhere is increased. The mediator of these compensatory changes is a chemical released by the adrenal medulla. It is known to pharmacologists as epinephrine and to the general public as adrenalin.

But adrenalin can do more than increase the synthesis and release of glucose. In a constellation of effects that psychologists call the fight or flight reaction, blood pressure and heart rate increase; we sweat, tremble, and feel nervous; we have an urgent need to go to the bathroom. (Recall the most terrifying, anxiety-provoking, or thrilling event of your life. Oddly enough, trembling with fear or with anticipated pleasure is mediated by the same chemical.) Fortunately, most of us most of the time release just enough adrenalin to help maintain appropriate levels of glucose, we experience no fight or flight effects, and in general we remain blissfully unaware of the complex and wonderfully orchestrated adjustments our bodies make in the service of our brains.

Fatigue, weakness, confusion, drowsiness, mental dullness, headache, heart palpitations, tremulousness, sweating, anxiety: what a marvelous set of symptoms; so ripe for misattribution. Those who preach that sugar is poison would have us believe that these common accompaniments of life are signs of neuroglycopenia or of the excess adrenalin release of hypoglycemia. The fact is that the vast majority of human beings is quite capable of adjusting via insulin, adrenalin, and other hormones to whatever carbohydrates are eaten. Furthermore, such adjustments are not pathological; they are the height of normality. Ignorance of these facts is almost universal among advocates of "health foods." More regrettable, many physicians are misguided as well. Patients who believe they suffer a disease called hypoglycemia (perhaps they've read *Sugar Blues*) are all too often encouraged in that belief by a physician who uncritically suggests or approves a low-carbohydrate "hypoglycemic diet."

A reliable means is needed to separate those rare individuals who have pathologically low levels of blood glucose from that very large number of people whose anxiety or fatigue or irritability has nothing to do with sugar. Many have been led to believe that an oral glucose tolerance test (OGTT) is that means. Following an overnight fast, a patient drinks a solution containing glucose. Blood levels of glucose are then measured for several hours. The OGTT is useful as a confirmation of diabetes and certain prediabetic conditions in which blood glucose remains elevated for extended periods. The OGTT is not a valid means to prove the presence of hypoglycemia. The reason is that the level of blood glucose that leads to symptoms of neuroglycopenia or adrenalin release is influenced by sex, age, and individual variation to such an extent that no predictions are possible.

It is not uncommon to find patients declared ill with hypoglycemia when blood sugar falls below 60 milligram-percent (mg-%) in an OGTT. Yet, a survey of 650 normal people by Arye Lev-Ran and Richard Anderson of the City of Hope National Medical Center revealed that 25% reached glucose levels below 54 mg-% and fully 10% dropped below 47 mg-%, all without symptoms of hypoglycemia. It is absolutely essential to the diagnosis of hypoglycemia that symptoms occur when glucose levels are lowest and that they disappear as glucose rises. When these criteria are applied, the disease of hypoglycemia is found to be relatively uncommon. This does not mean that fatigue, weakness, confusion, drowsiness, mental dullness, headache, heart palpitations, tremulousness, sweating, and anxiety are imagined or are to be ignored; simply that we should look elsewhere than to sugar for their causes.

If hypoglycemia is a "nondisease," what of the commonly held belief that refined sugar causes hyperactivity in children and violent behavior in adults? Regrettably, evidence to support these notions is very limited. I say regrettably because it would be wonderful if the home and school behavior of all, or even some, children called hyperactive could be improved by restricting sugar. It would be wonderful if violence in our society could be diminished and our prisons turned into shopping malls by a simple change in diet.

It is widely held that scientists ignore simple, "natural" answers to problems in favor of searching for complicated, esoteric solutions. This myth is perhaps an extension of the observation that many practitioners in the health industry value treatments in direct proportion to their monetary return. Thus we find drugs, surgery, and various forms of high technology used when simpler, less expensive means would do as well or better. This fact of life has little to do with arriving at simple or complex solutions to simple or complex problems of human health. Scientists do not ignore simplicity; they do demand evidence. It is that demand that often alienates those who believe, or wish to believe, that a simple answer is the correct answer.

*It is widely held that scientists ignore simple, "natural" answers to problems....
Scientists do not ignore simplicity; they do demand evidence. It is that demand that often alienates those who believe, or wish to believe, that a simple answer is the correct answer.*

Many parents are convinced that diet in general and sucrose in particular are leading causes of undesirable behavior in children. They have been supported in this view by the writings of physicians such as the late Benjamin Feingold (*Why Your Child Is Hyperactive,* 1973), William Crook (*Can Your Child Read? Is He Hyperactive?,* 1975), and Lendin Smith (*Food for Healthy Kids,* 1981)—all offer simple solutions to complex problems, and are presented by authority figures in skillfully marketed books. Sadly, our demand for evidence is usually answered by such good men as these not with scientific proof, but largely with testimonials and anecdotes that cannot by their nature serve in lieu of controlled scientific studies.

What of the studies that have systematically tested the relation between sucrose and hyperkinesis? I will tell you of just one, but its results are typical of many.

Dr. Mortimer Gross works at the Medical Center of the University of Illinois in Chicago. He has a longstanding interest in hyperkinetic children and each year treats about 150 of them. In the early 1980s, a 5-year-old boy came under his care who was "irritable, hyperactive, and distractable; he had a short attention span, visual and auditory perceptual deficits," and wet the bed nightly. His mother was convinced that her son's behavior was influenced by table sugar. Indeed, Dr. Gross tentatively agreed after testing the boy with lemonade sweetened with either sucrose, glucose, lactose, or saccharin; only after sucrose did the boy's behavior worsen.

Dr. Gross resolved to follow this observation up in other hyperkinetic children.

Over a period of two years, about 300 children were brought to Dr. Gross with the tentative diagnosis of hyperkinesis. Whenever a patient's mother stated that she was convinced that sugar caused or worsened her child's symptoms, mother and child were asked to participate in a study to support their belief. The mothers of 36 boys and 14 girls were given two quarts of lemonade, one sweetened with sucrose and the other with saccharin. The two containers were identified only by code. Thus, in the jargon of clinical trials, the mothers were "blind" to which contained sucrose and which saccharine. The mothers were instructed to give one-third of a quart from either container at a time when they would be able to observe their child for a few hours afterward. They were then to rate their child's behavior on a scale ranging from –5 (much worse) through 0 (no change) to +5 (much better).

When the results of the six trials in 50 children were assembled and the code broken, it was found that, in the words of Dr. Gross, "not one of the 50 children showed any consistent response to sucrose." What of the boy who was sensitive to sucrose? He and his mother moved away before more tests could be done. We will never know whether he was indeed affected by sucrose or whether Dr. Gross' initial observations were mere chance. From his overall experience, Dr. Gross concludes that "a hypersensitivity to sucrose can exist...(but) sucrose does not commonly affect hyperkinetic children adversely." A significant number of studies by other investigators in other places have reached the same conclusions.

George Moscone, the mayor of San Francisco, and Harvey Milk, a city supervisor, were shot to death in 1978 by Dan White. Mr. White was later convicted of manslaughter and in 1985 took his own life. He avoided a conviction for murder because of an imaginative defense and a gullible jury:

it was successfully argued that his violent behavior had been caused by a junk-food diet. This came to be known as the Twinkie Defense.

If any group has a vested interest in the idea that improper diet contributes to crime, it is the American Dietetic Association. After all it is the business of dietitians to give dietary advice and it would be very good for that business if criminals were simply malnourished. It is for this reason that the ADA's position paper of October 14, 1984 on diet and criminal behavior is of particular interest. The paper states that:

(a) Valid evidence is lacking to support the claim that diet is an important determinant in the development of violence and criminal behavior.

(b) Valid evidence is lacking to support the hypothesis that reactive hypoglycemia is a common cause of violent behavior.

(c) Inappropriate dietary treatment based on unfounded beliefs about the relationships between diet and criminal behavior can have harmful effects. It can:
   (1) Result in nutritional deficiencies and/or excesses.
   (2) Detract from efforts toward identification of effective treatment and prevention of the true causes of aberrant behavior.
   (3) Lead to the dangerous belief that diet, rather than the individual, has control over and responsibility for his behavior.

(d) A causal relationship between diet and crime has not been demonstrated. Diet is *not* an important determinant in the incidence of violent behavior. (ADA's emphasis)

Relatively few Americans are exposed to position papers of the American Dietetic Association. Millions read newspapers. The magazine section of my Sunday paper recently carried an article distributed throughout the country by

Knight News Service. The piece was written by Suzanne Dolezal, but might as well have been by William Dufty; Ms. Dolezal repeated much of what I believe is Dufty's nonsense. Entitled "The Sugar Overdose," there followed in 18-point type the statement that "The effects of hypoglycemia, low sugar in the blood from a high-sugar diet, may range from headache and depression to criminal acts." The continuing popularity of such writings seems verification of an opinion expressed 30 years ago that "Americans love hogwash."

Summing Up

The evidence that sugar, especially refined sugar, contributes to decay of our teeth is overwhelming. For that reason alone the use of concentrated and sticky forms of sucrose should be avoided, particularly by children and especially between meals. In contrast, we are nowhere near whelmed by the evidence that refined sugar is a specific causative factor in obesity, diabetes, heart disease,hypoglycemia, or behavioral disturbances. Diabetes and heart disease are associated with obesity—and if you drink too much alcohol, eat too much protein, fat, or carbohydrate, obesity will result. Diet certainly can influence behavior. But, with the exception of clear-cut states of deficiency, dietary effects are likely to be subtle. Attribution of all, most, or even a small fraction of behavioral disturbances in children, or violence in adults, to diet in general or sucrose in particular owes more to fiction than to science.

VITAMINS

CHAPTER 5

Vitamin A and Beta-Carotene

Of Carrots and Toxins

---

*A half million of the world's children
—wherever [they] are starving—
are rendered permanently blind each year
by [a] readily correctable deficiency disease....
a lack of vitamin A.*

An infant is brought to the visiting American physician. She has seen children like this before and is not surprised when the interpreter tells her that the baby is sick and doesn't open his eyes. A quick examination confirms her fears: the child is blind; behind his closed lids the corneas have dissolved into a shapeless jelly. The scene is repeated every day in southeast Asia, Africa, India, Brazil, Central America; wherever children are starving. The specific cause is a lack of vitamin A. A half million of the world's children are rendered permanently blind each year by this readily correctable deficiency disease.

Our knowledge of vitamin A did not begin with blind children, but with a more benign condition, the inability to see at night. Most have had the experience of entering a darkened movie theater and stumbling over those already seated. We have no problem upon leaving; those around us are clearly visible then. We have "gotten used to the dark."

75

Night blindness, the [eye's] failure to adapt to dim light, and its cure, juice squeezed from liver, were described 3500 years ago.....Hippocrates prescribed raw ox liver dipped in honey. Fishermen... discovered the efficacy of sea gull or codfish liver.... it remained for George Wald...to demonstrate that vitamin A is directly involved...For his achievements, Dr. Wald was awarded the Nobel Prize...

The process by which our eyes shift from seeing things in bright light to night vision requires the presence in the retina of a substance called visual purple or rhodopsin. Vitamin A (retinol) is an essential component of visual purple.

Night blindness, the failure to adapt to dim light, and its cure, juice squeezed from liver, were described 3500 years ago by an anonymous Egyptian physician. Many years later, but still before the birth of Christ, Hippocrates prescribed raw ox liver dipped in honey. Fishermen in various parts of the world discovered the efficacy of sea gull or codfish liver. Despite this practical knowledge, it remained for George Wald and his colleagues to demonstrate that vitamin A is directly involved in the synthesis of visual purple. Wald's investigations began in the early 1930s at the Kaiser Wilhelm Institute in Berlin and were completed in the Biological Laboratories of Harvard University. For his achievements, Dr. Wald was awarded the Nobel Prize for Medicine in 1967.

To explain the sightless child whom we met at the beginning of this chapter, we must consider night blindness as but the first in a series of ocular changes that occur when too little vitamin A is available. The second step is called xerophthalmia. Xero, as in Xerox, comes from the Greek word for dryness. Xerophthalmia begins as simple dryness of the eyes, but may quickly progress to softening of the cornea and total blindness. Accompanying these structural changes in the eye is a diminished resistance to infection; measles is a common cause of death among blinded children. Recent studies indicate that even a modest degree of deficiency may adversely affect immune function.

If in the 1980s hundreds of thousands of children are blinded by vitamin A deficiency, we might imagine that knowledge of xerophthalmia and its tragic consequences are new to medical science. This is hardly the case. On October 3, 1923, Dr. C. E. Bloch, Professor of Medicine at the University of Copenhagen, read a paper before the World's Dairy Congress in Washington, DC. He reviewed what was known prior to 1910 about xerophthalmia: epidemics in Brazil, Russia, and Japan; always among poor children and always in conjunction with a near-starvation diet. He told his listeners of xerophthalmia in animals. Finally he summarized his experience of a dozen years with poor children in Denmark.

A child taken from his mother's breast and fed little or no milk, butter, cream, fruits, or colored vegetables soon becomes listless and apathetic. Growth stops and weight loss begins. Infections of all kinds, but especially those of the skin and urinary tract, are likely. Night blindness progresses to xerophthalmia and eventual blindness if death from infection does not intervene. Yet, Dr. Bloch concluded, all of these effects are readily prevented by providing fresh milk, butter, or cod liver oil.

Dr. Bloch's remarkable presentation rested in large measure upon the work of Elmer Verner McCollum. Born in 1879 on a farm in Kansas, McCollum would become the most influential nutritionist of his time. After training in chemistry at the University of Kansas and at Yale University, he joined the faculty of the College of Agriculture of the University of Wisconsin in 1907. With a dozen rats purchased with his own money McCollum began to answer the question of why animals and humans cannot live on diets containing only protein, fat, carbohydrate, and minerals. Just six years

---

*...carotene is a "provitamin," that is...a precursor
substance...converted into vitamin A in the body.*

---

after his arrival in Wisconsin, McCollum together with Mar-
guerite Davis reported a growth-promoting factor in egg and
in butter; they called it "fat-soluble A."

Shortly after the appearance of the report by McCollum
and Davis, supporting evidence was published by Lafayette
Mendel, McCollum's teacher at Yale. But Mendel and his
colleague, Thomas Osborne, went further. They showed that
blindness in animals fed lard as the only source of fat could
be prevented by butter. The idea that blindness associated
with human starvation might have a similar cure was subsequently refined and focused by McCollum. He wrote in the
1918 edition of his book, *The Newer Knowledge of Nutrition*, that
xerophthalmia occurs in humans as a result of a specific lack
of fat-soluble A. A half-century later, none question that
conclusion, yet on the day that you read this sentence, several thousand children will be blinded for want of this substance.

For his part, McCollum was happy to call the anti-xero-
phthalmia factor "fat-soluble A." It neatly differentiated the
material from "water-soluble B," the anti-beri-beri factor
discussed in Chapter 6. But by this time, many influential
figures in nutrition had begun to accept the unifying notion
of Casimir Funk that there existed a whole new class of vital
nutritional factors. The three food factors for which substantial evidence then existed would thus be called vitamine
A, vitamine B, and vitamine C, the anti-scurvy factor. (The
"e" was later dropped from the word 'vitamine'.) Elmer
McCollum originally disliked the vitamin nomenclature, but
was persuaded to use the term in the 1922 edition of *The
Newer Knowledge of Nutrition*. Universal usage soon followed.

By 1920 all agreed that vitamin A is a fat-like material
found in things such as butter and egg yolk. Because of the
color of these foods and because of the inactivity of colorless

fats such as lard and almond oil, Harry Steenbock, a colleague of McCollum's at Wisconsin, proposed that vitamin A is carotene, the yellow pigment of butter and egg yolk. Others rejected the notion because some colorless fats had high vitamin A activity. The solution to this puzzle came with the discovery that carotene is a "provitamin,"that is, it is a precursor substance that is converted into vitamin A in the body. Conclusive proof was provided in 1929 by Thomas Moore in England. He showed that rats with no preformed vitamin A in their diets were able to maintain quite adequate levels of the vitamin when fed carotene.

We now know that there are more than 400 different carotenoids. They give us the yellow of egg yolk and butter and the orange of carrots, but also lobster-red, flamingo-pink, and salmon. Just as their hues are varied, so too is the pro-vitamin-A activity of the carotenoids. The most active is called "all-*trans*-beta-carotene," or simply beta-carotene, and the others are measured against it.

Beta-carotene is the predominant carotenoid in green leaves and in carrots, making each excellent foods for avoiding vitamin A deficiency. Upon eating your daily salad (a very good idea), beta-carotene passes to the intestine where some is absorbed unchanged and some is converted to a form of vitamin A. (As a part of their singular nature, cats are unable to cleave beta-carotene. In 1971, Donald McLaren and Beatrice Zekian of theAmerican University of Beirut described a 10-year-old Lebanese Arab girl who was cat-like with respect to beta-carotene and who like a cat required preformed vitamin A in her diet. This is surely a rare condition in humans.)

Upon eating your daily salad (a very good idea),
beta-carotene passes to the intestine
where some is absorbed unchanged
and some is converted to a form of vitamin A.

In 1989, the Food and Nutrition Board of the National Academy of Sciences set the Recommended Dietary Allowances (RDA) for vitamin A at 1000 retinol equivalents (RE) for adult males and 800 RE for adult women. During lactation, the amount is increased to 1200–1300 RE. Seems a simple matter. It isn't.

Even if we knew with some reasonable degree of precision the human need for vitamin A, which we don't, there are major uncertainties about the vitamin A activity of foods, especially those containing carotenes. The rate of metabolism of vitamin A and the rate of conversion of carotenes to vitamin A depend upon how much of the vitamin is already in the body. Other components of the diet are important; if too little fat is available, a common condition in areas of starvation, neither beta-carotene nor vitamin A will be efficiently absorbed; too little protein and the vitamin cannot effectively be mobilized from the liver. A final element of confusion arises because vitamin A is typically sold to American consumers not as retinol equivalents, but as International Units.

Few attempts have been made to determine the exact human need for vitamin A. An important reason for this is illustrated by a World War II study done in Sheffield, England. Sixteen volunteers were fed a vitamin A and carotene-deficient diet and observed closely for possible effects. After a month, a decrease in blood carotene levels was noted in all subjects. After eight months, blood retinol levels were significantly lower in half. After 11 months, three of the subjects developed night blindness. Others showed no signs of deficiency even after staying on the diet for more than two years.

A fact of life for all clinical investigators is that no two human beings are exactly alike in their response to a particular diet or drug or treatment. But the Sheffield results showed a degree of variation between subjects that is truly remarkable. The explanation lies in the fact that in times of plenty, our bodies set aside large amounts of vitamin A for future use. One estimate is that the typical American has

---

A fact of life for all clinical investigators is that no two human beings are exactly alike in their response to a particular diet or drug or treatment.

---

enough to last for two years, whereas some may be carrying around a 10-year supply. A happy situation for those about to be cut off from their usual supply of vitamin A, but a real problem for scientists who wish to study deficiency in normal volunteers in the laboratory.

The British work was far from a complete loss. Based upon the response of those few who developed night blindness and upon changes in blood levels when vitamin A was returned to the diet, it was estimated that the minimal protective amount was 390 retinol equivalents and a probably adequate intake would be 750 RE. These estimates provided a point of reference for all subsequent work and are not far from the present RDA of 1000 RE for the adult male.

Preformed vitamin A is found almost exclusively in animal tissues. Liver is by far the richest source; 3 oz. of beef liver provide about ten times the RDA. Because "liver and onions" is not a regular meal for most Americans, we may wish to consider the vitamin A content of eggs and milk. Each contains significant amounts, but each carries with it a penalty. As it comes from the cow, whole milk contains about 5% of the RDA per cup; it also contains saturated fat and, for most of us, unwanted calories. A single egg yolk provides 10% of the RDA of vitamin A; it also provides all the egg's cholesterol. We shall consider cholesterol and saturated fat in detail in Chapters 22 & 23, but for now let's assume that we don't want to increase our intake of either in our quest for vitamin A.

When all of the saturated fat is removed from whole milk, all of the vitamin A is removed as well. But read the label on a container of skim or low fat milk; each cup provides 10% of the RDA for vitamin A. This feat has been accomplished by adding back a larger amount of vitamin A than was originally present. As unnatural as it sounds, this is "food processing" and "food additives" at their best. An undesired element of the diet, saturated fat, has been removed and a desired one, vitamin A, has been provided in a palatable and widely consumed form.

Elsewhere in this book I have, and will, advocate a quart of skim milk per day for every adult. In addition to high quality protein, vitamin D, and calcium, that quart will provide 40% of the RDA for vitamin A. In fact that is about all of the preformed vitamin A that anyone needs. The rest of the requirement should come in the form of carotene. (Those of a vegetarian persuasion argue that in fact we don't need any preformed vitamin A; the body will convert carotene to vitamin A in optimal amounts. I don't dispute the fact of that conversion, but suggest that skim milk has other virtues as well and, with respect to vitamin A, will provide a desirable baseline intake for those less attracted to vegetables.)

The best guide to the carotene content of fruits and vegetables is color. A white potato contains none at all, whereas a sweet potato has a two-day supply. Green, yellow, orange, and red are sure to provide the provitamin, but in quite variable amounts. Those that give us the RDA or more in a single serving include broccoli, cabbage, chard, collards, kale, cantaloupe, pumpkin, mango, and spinach. (There's even some preformed vitamin A in spinach.) My personal choice is the carrot; eat even a small one each day and you needn't trouble yourself further about carotene. On cooked versus raw, there's little to suggest cooking is a problem but, as always, time, water, and temperature should be kept to a minimum.

In view of the abundance of carotene and vitamin A in the foods around us, it is remarkable that surveys suggest that the vitamin A intake of many Americans is deficient. Of greatest concern are children and the elderly, whose diets too often contain too little milk and too few vegetables.

The fact that some segments of the population have marginal intakes of carotene and vitamin A, the much publicized recent evidence of an anticancer effect of carotene or vitamin A (*see* Chapter 21), and the seemingly ineradicable notion that more is better, have led many to seek dietary supplements. The makers of such supplements have been only too willing to oblige. It's time to consider the toxicity of vitamin A.

Polar bears and humans have shared the Arctic for a very long time. The flesh of the bear has occasionally sustained humans, and on occasion, vice versa. From this extended association has come the sure knowledge among Eskimos and others of the region that the liver of the bear is poisonous and should not be eaten. European explorers have been rediscovering that fact on a regular basis for the past 400 years. Despite these many years of experience, it was not until 1943, 30 years after McCollum and Davis, that the toxic element of bear liver was identified as vitamin A.

In March of 1940 a three-year-old boy was brought to the Department of Pediatrics of the Johns Hopkins University School of Medicine. For the past 18 months he had had a poor appetite, been lethargic, and had lost much of the hair from his body. Hugh Josephs, the physician who examined him, found the boy's liver and spleen to be enlarged and the fingers and toes broadened and thickened. X-rays revealed abnormalities of the bones of the arms.

The boy had earlier received a diagnosis of a genetic disorder called Gaucher's disease, but Dr. Josephs suspected something else. In his words "the outstanding feature of the examination...was the enormously high content of vitamin A in the serum..." The mother then told him that from an age of two months, the boy had been given a teaspoonful of halibut liver oil every day and the child sometimes drank directly from the bottle. The teaspoonful alone would provide 240,000 IU of vitamin A (about 180 times the RDA for a child his age).

There are many other tales of human intoxication with vitamin A that could be told, but we will consider just one more. Sarah, a woman of 21 years, was admitted to a hospital in New York City on July 24, 1945. Her chief complaints were double vision, headache, and nausea. During her stay of a month and a half, no effective treatment was provided, but she was given a diagnosis: brain tumor.

Six days after her discharge from the first hospital Sarah entered a second. There she revealed that for the past two years, she had been taking vitamin A for her dry and scaly skin. She had begun with 25,000 IU a day, but soon increased the dose to 500,000, with an occasional day at a million units.

Among Sarah's examining physicians was a neurosurgeon. He thought her problems were caused by increased pressure of the fluid surrounding the brain. To relieve that pressure he removed a piece of her skull and drained off some of the fluid. After recovery from her surgery she left the hospital, but soon returned because chest pain and numbness in her pelvic area had been added to her previous problems. This time Sarah spent two months in the hospital. She was treated with thiamine and radiation therapy, but remained unimproved and undiagnosed. A neurologist favored the idea that a brain tumor was spreading.

Sarah entered her third hospital on February 10, 1947. There a neurosurgeon covered the previously made hole in her skull with a metal plate. Treatment then included "fever therapy"—intentional infection with typhoid fever, in an attempt to correct what had now been diagnosed as encephalitis, an inflammation of the brain. She was discharged on April 7th only to be admitted to still another hospital a month later.

Two symptoms were noted at hospital number 4: First, Sarah's pain was less severe if she remained motionless. To capitalize on that fact she was placed in a cast that ran from the nipple line to midthigh. Second, she was still taking 500,000 IU of vitamin A each day for her skin condition. Her physicians saw no reason to discontinue it. A neurologist suggested a possible viral infection of the nervous system.

Sarah entered her fifth hospital just five months after leaving the fourth. Pain in her joints had become so severe that she was able to walk only with great difficulty. A diagnosis of generalized infectious arthritis was soon made and she was discharged to the outpatient department for continued physiotherapy. For the next five years she was treated by chiropractic and osteopathic methods, continued to take her vitamin A, and lived in constant pain.

On February 5, 1953 Sarah came under the care of Alexander Gerber and his colleagues at the Jewish Hospital of Brooklyn. For the first time, the possibility of vitamin A poisoning was considered. All supplemental vitamin A was stopped. From that day her health began to improve. Within two months her skin texture was nearly normal and she had regained an appetite for food. For the first time in more than eight years she was free of pain.

Nearly a half century has passed since Sarah began her adventure with vitamin A. The lessons to be learned from her odyssey are no less fresh today:

(1) Vitamin A in excess is a poison.
(2) Intoxication with vitamin A can produce myriad effects: nausea, dry and peeling skin, pain in bones and joints are but a few. Increased pressure on the brain may cause signs and symptoms that, as in Sarah's case, are easily interpreted as the effects of brain tumor or infection.
(3) A general principle for both physicians and patients: Whenever a person becomes ill in mind or body for no apparent reason, every drug and dietary supplement should be evaluated as a possible cause.

It may appear reasonable to some readers for the Federal Government to step in and protect the public from this essential, but nonetheless toxic substance. The Food and Drug Administration in 1973 attempted to do just that. It was proposed that all vitamin and mineral supplements

> *Though there is no general agreement on what*
> *daily intake of vitamin A will get us into trouble,*
> *it's hard to argue with the opinion expressed in*
> *1980 by the...National Academy of Sciences that*
> *anything in excess of 25,000 IU is "not prudent."*
> *I'd go a little further and opine it's stupid.*

containing more than 150% of the recommended daily allowance be classified as drugs and hence subject to regulation. The courts wouldn't permit it. The FDA next focused on vitamin A. A regulation was published requiring a physicians's prescription for any daily dose in excess of 10,000 international units. In 1978, the FDA lost again in court. We retained our right to poison ourselves with vitamin A.

The Food and Nutrition Board of the National Academy of Sciences has chosen to express their Recommended Dietary Allowance in terms of retinol equivalents (RE). The RDA for an adult male is thus 1000 RE. Confusion is introduced by the fact that sellers of vitamin supplements almost always express the content of their pills not in RE, but in International Units (IU). A general purpose multivitamin typically contains 5000 IU and this is said to be 100% of the RDA. That statement is usually wrong. The reason is that we must consider the source of the "IUs." The RDA is based on the assumption that one-half of the requirement will be met by preformed vitamin A and one-half by beta-carotene. Thus, 5000 IU from fish liver oil equals 1500 RE or 150% of the RDA, and by contrast, 5000 IU from beta-carotene equals 500 RE, or 50% of the RDA. So long as we remain rational in our use of supplements, no harm is likely to be done by erroneous conversion of international units of vitamin A or beta-carotene to retinol equivalents. Unfortunately, however, rationality is in short supply in the typical vitamin store or catalog. A common "Vitamin A" supplement found there contains 25,000 IU from fish liver oil. Converted to retinol equivalents, that's 750% of the RDA. Though there is no general agreement on what daily intake of vitamin A will get us into trouble, it's hard to argue with the opinion expressed in 1980 by the Committee on Dietary Allowances of the National Academy of Sciences that anything in excess of 25,000 IU is "not prudent." I'd go a little further and opine it's stupid.

"Teratology" is an uncommon word; the Greek word, teras, means monster; teratology is the science of malformations. In 1953, S. Q. Cohlan reported that rats given large amounts of vitamin A gave birth to malformed babies. Since then, the teratogenic effects of vitamin A have been proven in several species and a dozen case reports have linked excess vitamin A intake with human malformations. Only recently have epidemiological data become available, but these too suggest that vitamin A is a teratogenic risk for humans.

The concern that unborn babies may be harmed by vitamin A taken by their mothers is intensified by what is known about isotretinoin, an anti-acne drug sold in this country with the trade name Accutane. Isotretinoin is very close in chemical structure to vitamin A, and in animals produces a range of malformations very similar to those caused by vitamin A. It is now well established that isotretinoin use by pregnant women can cause birth defects. Fear that unborn babies might be harmed by excessive intake of vitamin A by their mothers is one of several factors that have led some authorities to suggest that the RDA be decreased.

For pregnant women and for the rest of us, a reasonable alternative to dietary supplementation with preformed vitamin A is to use beta-carotene. The human body is well able to control the conversion of beta-carotene to vitamin A and there is no instance known where vitamin A intoxication occurred as a result of taking too much carotene. Furthermore, no human birth defects have been associated with the use of beta-carotene. In a position paper issued in 1987, the Teratology Society, a group devoted to the study of malformations,

---

*If a vitamin supplement is used, keep...preformed
vitamin A at 5000 IU or less and limit yourself to
one tablet per day. Better yet, stick to beta-carotene.
Resist the idea that if a little is good, more is better.*

---

suggested that "beta-carotene be considered the primary
source of (vitamin A) for women in their reproductive years..."

One other thing must be said about beta-carotene. Taken in massive amounts it will cause those of the white race to turn yellow, those of the yellow race to grow somewhat orange, and the soles and palms of blacks to become yellowed. This is demonstrated each year by a few people who take to drinking carrot juice by the quart or eating tomatoes by the gross. (A product called Orobronze contains canthaxanthin, a carotenoid reputed to impart a suntan-like hue.)

Summing Up

Although deficiency of vitamin A is a leading cause of blindness in less affluent nations, it is an excess of the vitamin that poses the greater hazard to Americans. True, not many of us are likely to be poisoned by too much polar bear liver; but potentially toxic amounts of the vitamin can be found in every vitamin store and catalog. It should be obvious to readers of this chapter that the human need for vitamin A is readily met by the carotenes contained in fruits and vegetables. In addition, all but strict vegetarians obtain preformed vitamin A in eggs, milk, and other dairy products. Liver and liver oils from cattle, poultry, and fish are not a necessary part of the diet. Indeed, a few infants have been poisoned by mothers convinced that chicken liver is a great food for growing babies. If a vitamin supplement is used, keep the content of preformed vitamin A at 5000 IU or less and limit yourself to one tablet per day. Better yet, stick to beta-carotene. Resist the idea that if a little is good, more is better.

CHAPTER 6

Thiamine (Vitamin B₁)

Polished Rice and Beriberi

---

*Beriberi is a disease that takes its name
from the Singhalese word for weakness....
"Picture yourself a skeleton, with a parchment-like
wrinkled skin...drawn over it. A weary, dilapidated
individual, apparently a picture of misery,
with a staff to assist his tottering footsteps
if the power of locomotion still remains to him."*

Beriberi is a disease that takes its name from the Singhalese word for weakness. Percy Netterville Gerrard, district surgeon of the Federated Malay States civil service, gave us in 1904 the following description. "Picture yourself a skeleton, with a parchment-like wrinkled skin...drawn over it. A weary, dilapidated individual, apparently a picture of misery, with a staff to assist his tottering footsteps if the power of locomotion still remains to him." Later workers would also note a variety of behavioral disturbances: decreased attention span, personality changes, depression, lack of initiative, and poor memory. (Colonials readily interpreted the more subtle effects of beriberi as a reflection of inborn characteristics of the natives.)

Under various names, beriberi has been known in the Orient for several thousand years. In Japan it was called kakke, a rare condition found almost exclusively among the wealthy. That exclusiveness began to erode in the 17th century, with epidemics occurring in Tokyo and other cities. By 1870 beriberi was a nationwide problem whose incidence would continue to increase for the next 50 years. Ironically, it was an officer of the Japanese Naval Medical Service who had provided clear evidence in the 1880s of a nutritional cure for beriberi.

The Japanese warship Riujo sailed from Japan in 1882 with a crew of 276. After stops in New Zealand and Chile, she arrived in Honolulu. During the voyage of 272 days, 60% of the men suffered from beriberi and 25 died of the disease.

Dr. Kanehiro Takaki was in 1882 a junior medical officer recently returned from five years of study in England. He had observed that beriberi was rare in the Royal Navy despite voyages of equal length and generally comparable living conditions. In seeking differences between the British and Japanese sailors, Takaki focused on their respective diets. The British consumed large amounts of animal protein, whereas the principal component of the Japanese diet was polished rice. (When separated from the leaves and stalks of the plant, a grain of rice remains encased in its indigestible hull. Hand grinding will remove the hull, but the pericarp or "bran coat" is left behind. The resulting "brown rice" is edible, but becomes rancid in storage. When steam-driven milling machines became available, large quantities of rice could be stripped of their bran coats and buffed to produce "polished rice.")

Dr. Takaki hypothesized that beriberi was caused by a deficiency of nitrogen in the diet. (Protein is the only significant dietary source of nitrogen.) This hypothesis was tested in 1884 when the warship Tsukuba retraced the path of the ill-fated Riujo. At Takaki's insistence, Tsukuba carried with her ample supplies of meat and dried milk. By voyage's end, none of the crew had died of beriberi and signs of the disease had been seen in only 14. Meat and dried milk became

> *Beriberi remained an uncommon scourge*
> *of the rich so long as only they could afford the*
> *lovely whiteness of polished rice. Technology in the*
> *form of steam-driven milling machines brought*
> *polished rice, and beriberi, to the common man.*

a part of the diet of the Japanese sailor and beriberi was virtually eliminated from the Emperor's fleets. Non-oceangoing Japanese were less fortunate.

Despite Takaki's success in preventing beriberi by dietary means, others must be credited with recognizing the true nature of the disease and for providing a specific remedy. For several hundred years, the Netherlands controlled a vast area of the Pacific, the Dutch East Indies. As a result, Dutch medical officers had ample opportunity to study beriberi in the prisons and insane asylums of Kuala Lampur and Djakarta and elsewhere. By the 1880s, their interest centered on the influence of rice. It was known that prisoners or patients fed polished rice were 250 times as likely to get beriberi as were those fed brown rice. Christiaan Eijkman, a Dutch physician on the island of Java, showed in 1897 that a beriberi-like condition occurred in chickens fed polished rice. Furthermore, the chickens were cured by feeding them a water extract of brown rice. Eijkman thought that a toxin in rice was being unmasked in the conversion of brown to white rice. (For his work, Eijkman shared the Nobel Prize in Medicine for 1929 with Sir Frederick Gowland Hopkins, a British pioneer in the study of vitamins.)

It remained for Eijkman's successor on Java, Gerritt Grijns, to suggest the true cause of beriberi. He proposed in 1909 that an essential substance was being removed from rice during the milling process and that it was a deficiency of this substance that caused beriberi in people as well as chickens. Furthermore, the essential substance was not confined to rice;

beriberi was also prevented by water extracts of green peas, beans, and meat.

The Japanese experience could now be explained. Beriberi remained an uncommon scourge of the rich so long as only they could afford the lovely whiteness of polished rice. Technology in the form of steam-driven milling machines brought polished rice, and beriberi, to the common man. Takaki's achievement in ridding the Japanese navy of beriberi did not result from the introduction of more animal protein into the diet, but to the water-soluble substance that happened to come along with the protein.

The protective substance found in rice and other foods had come to be called the "antineuritic" factor, a reference to the neuritis of beriberi in which both sensory and motor functions are impaired. To distinguish it from the previously discovered "fat-soluble A," some referred to it as "water-soluble B." A more euphonious name was needed and Casimir Funk provided it. Funk realized in 1911 that water-soluble B belonged, together with fat-soluble A and water-soluble C, to an entirely new class of food factors. He named this class "vitamine," a contraction of "vital amine." The "e" was dropped when it became apparent that not all members of the class were amines. Water-soluble B thus became vitamin B and, when other water-soluble vitamins were found in the same sources, it was designated vitamin B_1. Finally, with the identification of its chemical structure, the official name, thiamine, was given. Today, vitamin B_1 and thiamine refer to the same chemical substance.

We still occasionally find someone referring to thiamine as antineuritic factor. Confusion may result. While a student at the Culinary Institute of America, my elder daughter showed me one of her textbooks, the English translation of Prosper Montagne's *Nouveau Larousse Gastronomique*, subtitled "The World's Standard Encyclopedia of Food, Wine, and Cookery." I suspect that one could gain weight just looking at the pictures. Anyway, the book describes vitamin

B₁ as "an antineurotic factor" and goes on to say that "absence of vitamin B from the diet gives rise to neurotic complaints." Though the words "neuritic" and "neurotic" differ by but a single letter, their meanings and implications are far different. I can't help wondering how many gourmets, professional chefs, and other readers of Larousse are now treating their neuroses with thiamine. (I've never seen the French edition of the book, so I don't know whether the error was by the author or by the translators.)

The Recommended Dietary Allowance for thiamine varies with age, sex, and caloric intake, but we can use 1.6 mg as a convenient reference point. That is the maximum allowed for anyone, and happens to be for nursing mothers. The minimum amount for adults of either sex and any age is about 1.0 mg per day. How far below these recommendations one can drop before getting into trouble isn't known with certainty.

Because the signs of beriberi include loss of appetite, constipation, and fatigue, many tired, constipated people with poor appetite have been led to believe they need more thiamine. This is rarely the case. Though it's often been suggested that "marginal" thiamine deficiency is common in the United States, a number of surveys have failed to support the idea. The probable reason for our relatively good thiamine status is the nearly universal practice of "fortifying" flour with thiamine. Check the label of Wonder Bread, one of those pure white homogeneous products so abhorred by advocates of natural foods, and you'll find that it provides as much thiamine as any whole-grained bread. This is a case in which a "food additive" isn't such a bad thing. (This is not meant as an endorsement of the practice of "refining and fortifying" in which some things are taken out of a food and a few are put back in. More is lost in the refining of cereal grains than thiamine.)

The classic descriptions of beriberi provided us by the Dutch and British physicians of the Malay Archipelago were surely contaminated by the effects of multiple vitamin and macronutrient deficiencies. However, there is no reason to doubt that the major signs of beriberi are caused by lack of thiamine. In a Japanese investigation done in 1937, otherwise well-fed subjects were limited to 0.3 mg of thiamine per day. Clearcut signs of beriberi appeared after about 3 months. If thiamine intake is reduced to near zero, effects may be seen much more rapidly; another Japanese study detected changes in heart function after only a week. This is consistent with the fact that thiamine, like the other water-soluble vitamins, is stored in the body in only limited amounts.

Four forms of thiamine deficiency disease are generally recognized. The first type, infantile beriberi, occurs in babies nursed by thiamine-deficient mothers. It is a frequent cause of death in areas of the world where gross malnutrition is common. The person described by Gerrard in the opening paragraph of this chapter ("...dilapidated...picture of misery...with...tottering footsteps...") suffered from the second variety of the disease, which was then and is still called "dry" beriberi. When thiamine deficiency is somewhat more profound, the third manifestation of the disease, "wet" beriberi results, taking its name from the accumulation of fluid in the body; Gerrard asks us to "imagine a patient swollen beyond recognition." Finally, there is a fourth form of thiamine deficiency that is seen exclusively in alcoholics: the Wernicke-Korsikoff syndrome. Alcoholics often carry the double burden of an inadequate diet and alcohol-induced impairment of thiamine absorption. Although some of the aspects of Wernicke-Korsikoff syndrome respond favorably to treatment with thiamine, others seem to reflect permanent damage to the brain. Fortification of alcoholic beverages with thiamine to prevent Wernicke-Korsikoff syndrome has often been suggested but, to my knowledge, never been implemented.

The major traditional sources of dietary thiamine are the bran coats of cereal grains, fresh vegetables, especially green beans and peas, pork, beef, and organ meats. In addition, however, many processed foods have had thiamine added to them. I've already mentioned the fortification of flour and, if you read the labels on typical breakfast cereals, you'll find lots of thiamine there as well.

If you consume in a day a quart of milk, two slices of bread, and two ounces of Cheerios or Corn Flakes or any of the presweetened products aimed at children, you will obtain the recommended allowance for thiamine; and that's without eating any vegetables or meat at all. With that example in mind, it's a little hard to see how we need to buy brewer's yeast or wheat germ or blackstrap molasses in order to get enough thiamine. (I chose those as "sources of vitamin B$_1$" listed in a "vitamin and mineral guide" found in a "natural food" store.) We most certainly don't need 100, 250, or 500 mg tablets of vitamin B$_1$.

If my diet provides the RDA of thiamine, 99.7% of a 500 mg tablet goes straight down the toilet. Even with depletion to the point of beriberi, the body is able to absorb only about 10 mg per day.

Despite the ease with which we can meet the RDA for thiamine, we should keep in mind that the vitamin can be lost from foods during cooking. Though we might imagine that thiamine would be destroyed by heat, the major culprit seems to be water. Thiamine-enriched flours are little affected by baking or toasting, but significant losses can occur when spaghetti or macaroni are cooked in water; the vitamin simply dissolves in the water and is drained off. A study done at the American Institute of Baking found losses of about 50%. Even with that, a 12-oz serving of spaghetti still provides 35–40% of the RDA for thiamine. The general rules to be followed are to use a minimum amount of water, to cook no longer than is necessary, and to make use of as much of the cooking water as is possible.

Some foods contain substances that either inactivate thiamine or block its actions. This interesting bit of information is sometimes brought to public attention by those with vitamins to sell. In fact, we have little reason to be worried.

---

*Those of us who consume even a modestly varied diet
have no difficulty in meeting the RDA for thiamine.*

In some fish there is an enzyme that destroys thiamine and,
if I were on a steady diet of unfortified polished rice and raw
fish, I'd probably be in trouble. But cooking inactivates the
enzyme and, besides, raw fish has never been one of my pas-
sions. There's no question that tea and coffee contain small
amounts of anti-thiamine chemicals, but again, they are of
doubtful importance in human nutrition. About 10 years ago,
a study of these thiamine blockers was done in Thailand. A
group of natives was examined who chew tea leaves all day
and who drink large amounts of tea as well, presumably for
the stimulatory effect of caffeine. Thiamine levels were only
slightly lower than normal and no signs of thiamine deficiency were found.

The fact that 500 mg tablets of thiamine are for sale leads
me to believe that some people take enormous amounts of
the vitamin. Fortunately, no harm is likely to be done. The
only reports of thiamine toxicity followed its injection, and
in those cases it's hard to determine the source of the problem.

Summing Up

For many centuries in much of the world, the primary
source of thiamine was brown rice. Only after technology
permitted the 'polishing' of rice to remove its bran coat did
beriberi, the classic disease of thiamine deficiency, become
widespread. In this country at this time, severe thiamine
deficiency is seen primarily in alcoholics. Untreated, they may
suffer irreversible brain damage. Those of us who consume
even a modestly varied diet have no difficulty in meeting
the RDA for thiamine. Rich sources include whole grains,
fortified flour and its products, green vegetables, pork, beef,
and milk.

CHAPTER 7

Niacin

Pellagra and Madness

Columbus could not have imagined that...
[by] introducing corn into Europe he would lay
the groundwork for pellagra, a disease that would
spread across the Continent and would,
[even in the 20th] century, kill tens of thousands
of poor whites and blacks in the United States.

When Christopher Columbus reached the West Indies, corn had been a part of the diet of Native Americans for at least 5000 years. Columbus could not have imagined that by introducing corn into Europe he would lay the groundwork for pellagra, a disease that would spread across the Continent and would even, in the first third of this century, kill tens of thousands of poor whites and blacks in the United States.

The characteristics of pellagra are usually remembered as the three Ds: dermatitis, diarrhea, and dementia. In fact, the disease is much more complex than that. Early signs such as weakness and lack of energy are easily confused with laziness and deficiency of character, a not uncommon interpretation when they appeared in mill workers and tenant farmers. Writing in 1940, J. P. Frostig and T. D. Spies said: "Men previously strong, courageous, and enduring become shaky, weary, and apprehensive even before the generally recognized signs of pellagra develop...(it) breaks down the morale..." Insomnia, depression, and impaired memory of-

---

The characteristics of pellagra are usually
remembered as the three Ds: dermatitis, diarrhea,
and dementia....It takes its name from the Italian
words for "rough skin." That is too mild a term.
The skin lesions of pellagra are terrible to behold...

---

ten progressed to delusions, hallucinations, and dementia. Twenty-five percent of the patients admitted in 1910 to an insane asylum in South Carolina were found to have pellagra.

Unlike beriberi, pellagra is not an ancient disease. It was first observed in 1735 by Gaspar Casal, physician to Philip V of Spain. It takes its name from the Italian words for "rough skin." That is too mild a term. The skin lesions of pellagra are terrible to behold; I particularly recall the photograph of a girl, perhaps seven years of age, with a ribbon in her neatly combed hair, whose hands and face were swollen and cracked by pellagra. As the disease progresses, the effects upon the nervous system become more obvious, with irreversible insanity the ultimate phase.

The cause of pellagra eluded scientists for hundreds of years, but its association with corn was evident to all. Although Casal suspected that pellagra was caused by faulty nutrition, most theories of the 18th and 19th centuries suggested a toxin or germ in spoiled corn. The public health officials of France of the 19th century took a pragmatic approach; they discouraged the eating of corn and the disease virtually disappeared in their country.

In formulating his concept of "vitamines" in 1911, Casimir Funk had proposed that scurvy, beriberi, rickets, and pellagra were probably deficiency diseases. We have seen in the last chapter how water-soluble vitamin B had been shown to have anti-beriberi properties and, for a time, many believed the matter of vitamin B to be settled. However, in the constant ferment that is science, evidence began to accu-

mulate that water-soluble vitamin B was not a single substance, but a group of substances, the B complex. Thus began
what Leslie Harris has called "this long history of disentan-
glement"; it would eventually yield thiamine (B_1), niacin,
riboflavin, pyridoxine, pantothenic acid, and biotin.

Although pellagra probably was present in the United
States in the early 1800s, and surely occurred in Civil War
prisoners, it was not generally known to the medical profession until 1907, when George Searcy described 88 cases in an
Alabama insane asylum. So long as only those living in institutions, mainly the orphans and the insane, were affected
by pellagra, not much note would be taken of it. However,
within two years of Searcy's report on pellagra, cases had
been identified among workers in more than 20 states. By
1914, pellagra was epidemic in the American South, particularly among sharecroppers and in the textile mills. Political
pressure to do something about it became irresistible.

In March of 1914, Joseph Goldberger was a 40-year-old
physician in the United States Public Health Service. When
the Surgeon General assigned him to study pellagra, he was
already a 15-year veteran of investigations into measles,
parasitic worms, diphtheria, malaria, typhus, typhoid, den-
gue, and yellow fever. Although the cause of pellagra was
obscure, most American authorities believed it to be an infectious disease. Given Goldberger's past experience with
insect carriers of disease and the prevailing belief in microorganisms as the cause of pellagra, his natural inclination
might have been along those lines. His genius is revealed by
the fact that three months after beginning his investigations,
he published a paper in which he concluded that pellagra is not
infectious, the cause is dietary, and it can be prevented by an
increase in the consumption of fresh meat, eggs, and milk.

Goldberger's hypothesis met considerable resistance.
The Pellagra Commission of the State of Illinois had concluded just three years earlier that pellagra was caused by
infection with a microorganism and, in 1912, the Thompson-

McFadden Pellagra Commission in South Carolina had reached the same conclusion.

Goldberger knew that his hypothesis must be tested. He did so in three ways:

(1) The children of an orphanage in Jackson, Mississippi received additional meat, eggs, milk, beans, and peas for a period of two years. Pellagra, which had been rampant, disappeared. When the money ran out and the children's diets returned to "normal," the disease came back.

(2) He put a group of prisoners at the Rankin Farm of the Mississippi State Penitentiary on a diet like that of the pellagrans; signs of the disease appeared after three months.

(3) He tested the infectious nature of pellagra in himself, his wife, and fourteen of his colleagues. They injected themselves with the blood of pellagrans; they swallowed capsules containing nasal secretions, bits of dead skin, urine, and feces of pellagrans; they did not catch pellagra.

By the end of 1915, Goldberger and his associates had demonstrated "the complete prevention of pellagra by diet alone." We might imagine that pellagra ceased from that time to be a public health hazard. It did not. In 1916, pellagra was the second leading cause of death in South Carolina. During the years 1924 through 1939, no fewer than 2000 and as many as 8000 Americans died of the disease. W. Henry Sebrell, one of Goldberger's early associates, estimates that there were 250,000 cases in 1928.

Why did pellagra continue to kill Americans long after Goldberger demonstrated its dietary basis? Some would argue that the application of all advances in science requires the passage of time to win over those who hold other beliefs. Leslie J. Harris, former Director of the Dunn Nutritional Laboratory of the University of Cambridge thought otherwise: "The explanation is economic, not scientific...It is a sad commentary that while these people were dying from a dietary deficiency, at the same time in other parts of the country food was being burned or thrown into the sea because of 'over-production'." It is a sadder commentary that more than 30 years after Harris wrote those words, we in the United States still find ourselves uncertain as to the extent of "hunger in America"; in the midst of enormous stockpiles of food, we are still confronted with malnutrition in significant numbers of children, pregnant women, and old people.

As far as Goldberger was concerned, there remained only one question: what was the dietary factor that was preventing and curing pellagra? Oddly enough the search for that factor was aided immensely by a disease called blacktongue. Veterinarians in the rural South had been aware of blacktongue in dogs for many years, but it was not until 1916 that it was suggested that the disease was analogous to human pellagra. Impressed by the fact that the occurrence of blacktongue paralleled that of pellagra, Goldberger soon showed that the disease could be produced at will by feeding dogs the diet associated with pellagra. Goldberger was no longer dependent upon long and difficult experiments in orphans and the poor; the antipellagra factor could now be investigated in dogs. Despite this advantage, progress was slow.

In 1922, after eight years of work, the conclusion was reached that pellagra was caused by an amino acid deficiency; of the known vitamins, A, B, and C, none cured it, whereas protein-rich foods such as milk and beef did. That hypothesis was soon abandoned when it became clear that not all proteins were effective, though protein-poor dried yeast was. Finally, in 1925, Goldberger and his associates suggested that "...there is a heretofore unrecognized or appreciated dietary factor which we designate as factor P-P," the pellagra-preventative factor. It soon became clear that factor P-P was in fact a new vitamin; the British called it vitamin B_2 to distinguish it from the anti-beriberi factor, vitamin B_1. In 1929, the year of Goldberger's death, the American Society of Biologi-

---

*Confusion of nicotinic acid with nicotine,
the alkaloid of tobacco, was inevitable.
To remedy the situation, the Food and Nutrition Board
suggested a new name, niacin.*

---

cal Chemists honored him by naming the pellagra-preventative factor, vitamin G.

Nicotinic acid had been known to chemists since 1897; it takes its name from the fact that it shares a part of the nicotine molecule. Funk had isolated nicotinic acid from rice polishings but, because it was ineffective against beriberi, little note was taken of it. It was not until 1936 that nicotinic acid was shown to be important to the function of several enzymes and to be required for the growth of a number of microorganisms. Only then did interest arise in the possibility of a nutritional role for nicotinic acid. The crucial experiment was done in 1937 by R. J. Madden, a student of Conrad Elvehjem's at the University of Wisconsin. He cured black-tongue in dogs with nicotinic acid and, on November 5, 1937, the cure of human pellagra by nicotinic acid was reported at the Central Society for Clinical Research in Chicago. Goldberger's pellagra-preventative factor was nicotinic acid.

The need for another name for nicotinic acid soon became evident. When the proposal was made in 1939 to fortify bread with nicotinic acid, a headline read "Tobacco in Your Bread." Confusion of nicotinic acid with nicotine, the alkaloid of tobacco, was inevitable. To remedy the situation, the Food and Nutrition Board suggested a new name, niacin.

Despite the success of niacin in curing pellagra, there were still a number of puzzling aspects. If the pellagra-preventative factor is niacin, how are niacin-poor foods such as milk able to prevent pellagra? This is the way that Conrad Elvehjem put the question to Willard Krehl, another of his graduate students: "Why is it that milk, which contains very little niacin, effectively cures and prevents pellagra in man and blacktongue in dogs, while corn, which is very much richer in niacin, apparently is a major factor in the production, both of pellagra in man and blacktongue in dogs?" As would any good scientist, Krehl answered the question not with words, but with an experiment. He fed tryptophan, an essential amino acid found in abundance in meat and milk, to dogs maintained on a pellagrous diet; they grew normally and remained healthy. Tryptophan must therefore somehow be able to substitute for niacin. Later work would show that some of the dietary tryptophan is converted to niacin. Pellagra might as well be called a disease of tryptophan deficiency.

A second puzzle: why did pellagra not appear in Native Americans? In Mexico, for example, all of the conditions for pellagra seemed present: poverty, lack of animal protein, a corn-based diet. The answer seems to lie in the manner of preparation of the corn. Since prehistory, these people have soaked their corn in lime water for extended periods before eating it. This process converts niacin in the corn into a form that is readily absorbed by the human body. Here again we have an example of "food processing" that adds to the nutritional value of the food. Had the peoples who adopted corn as a part of their diet adherred to this method of processing, pellagra might not have appeared at all.

In deciding whether our diets contain enough niacin, we must not forget that tryptophan is converted into niacin in an approximate milligram ratio of 60 to 1. Thus, our accounting must include tryptophan as well as pre-formed niacin. It is for this reason that the present RDAs speak of "niacin equivalents." A niacin equivalent is equal to 1 mg of niacin or 60 mg of tryptophan.

For reasons unknown to me, niacin equivalents are not used on all food labels. This could seriously mislead people about their niacin status. For example, a one-cup serving of milk is said by the label to contain less than 2% of the RDA for niacin; an undeniable fact. However, a cup of milk does

contain 1.8 niacin equivalents, largely in the form of tryptophan. A quart of milk provides about 50% of the RDA for
niacin equivalents, yet the label would have us believe that
milk is a poor source of the vitamin.

Another possible point of confusion with respect to niacin is the closely related substance, nicotinamide (nicotinic
acid amide, niacinamide). In our bodies, niacin is converted
into nicotinamide and is used in that form. In terms of sup-
plying the RDA, niacin and nicotinamide are virtually inter-
changable on a milligram basis; 20 mg of nicotinamide is as
good as 20 mg of niacin. However, there are important differences between the two. In doses of several hundred milligrams
or more, niacin can cause flushing of the skin, intense itching,
stomach upset, and a variety of other adverse effects. Nicotinamide is much less toxic and this is probably the principal
reason why it has been favored by the health food stores.

No exact statement can be made about the daily need
for niacin. The current RDAs are based upon studies done
in human volunteers in the early 1950s by Grace Goldsmith
at the Nutrition Research Laboratory of Tulane University
and by M. K. Horwitt at Elgin State Hospital in Illinois. Their
results were not in complete agreement, but it certainly seems
safe to conclude that the recommended intake of between 13
and 20 niacin equivalents per day for adults is quite adequate.

Significant amounts of preformed niacin are found in
beef, pork, chicken, and many other meats. For example, a
quarter pound of ground beef contains about 5 mg of niacin,
and an additional five niacin equivalents in the form of tryptophan. Smaller amounts of preformed niacin are found in a
wide variety of fruits and vegetables. Since 1939, virtually
all flour used in the United States has been fortified with niacin; two slices of most breads provide 2–4 mg. Nearly all
breakfast cereals are now supplemented with niacin at a level of 5 mg per ounce. If we now add in the niacin equivalents from tryptophan, it's not hard to see why pellagra has

become a rare disease in America. An adult who consumes 58 grams of mixed animal and vegetable protein (the RDA) gets about 12 niacin equivalents from that source alone. Finally, coffee, a much maligned component of many diets, contains 1–3 mg of niacin per cup.

The many sources of niacin equivalents in the typical diet, together with the fact that niacin is quite resistant to destruction by cooking or in storage, make it a simple matter to insure an adequate intake of the vitamin. Goldberger had no trouble eliminating the disease in his orphans simply by adding meat and milk to their fare. Despite these facts, our friendly health food store is pleased to offer us still more of the vitamin, usually in the form of nicotinamide (they usually call it niacinamide, sometimes "vitamin B_3"). The nicotinamide may be part of a multivitamin, a component of a "B Complex," or sold by itself. One recent catalog offers "niacinamide" in 100, 250, and 500 mg tablets; never mind that 500 mg is at least 25 times the amount any body can use.

Until quite recently there wasn't much to be said about uses for niacin other than its recognized value in correcting niacin deficiency. That situation changed abruptly with the publication in 1987 of an enormously popular book by Robert Kowalski. In *The 8-Week Cholesterol Cure*, Mr. Kowalski suggests that those with an interest in reducing their level of serum cholesterol should eat lots of oat bran and take massive doses of niacin. We'll come back to the "cure" in Chapter 22. For now, I'll just give you the background of Mr. Kowalski's advice regarding niacin.

The title of this chapter connects pellagra and madness. It is thus appropriate that the discovery of niacin's anticholesterol effects arose from studies in a department of psychiatry. In the early 1950s, Abraham Hoffer and Humphry Osmond at the University of Saskatchewan were studying niacin as a treatment for schizophrenia. Their colleague, Rudolf Altschul in the Laboratory of Gerontology, was interested in ways to decrease serum cholesterol levels. In a

---

> *...niacin has been promoted...as a do-it-yourself
> cure for elevated cholesterol; it is not.*

---

collaborative study, Drs. Altschul, Hoffer, and J. D. Stephen found that medical students given 4000 mg of niacin over a period of 24 hours experienced an 8% fall in cholesterol.

The status of niacin as a vitamin and its ability in minute quantities to prevent and to cure pellagra are without question.   In contrast, niacin's value as a drug in massive doses to treat mental illness and to reduce cholesterol levels remains to be established. It took about 15 years for the majority of the psychiatric community to decide that niacin is useless in schizophrenia; I am sure there are some who yet favor its use. In contrast, many studies confirmed the cholesterol-reducing properties of niacin. Unfortunately, it is for many a distinctly unpleasant drug and it soon fell into disuse. Today's renewed interest can be attributed to the cholesterol frenzy of the 1980s and to the realization that other drugs to lower cholesterol may be even worse than niacin.

Summing Up

Niacin is both a vitamin and a drug. In the small amounts specified by the RDA, niacin, the vitamin, prevents the development of the deficiency disease called pellagra. In the American South of the 1920s and 30s, pellagra often occurred among the poor who depended upon corn as their primary source of protein. Pellagra is now uncommon in the United States, in large measure because of the fortification of flour with niacin. In doses hundreds of times greater than the RDA, niacin is a drug that may have a role in controlling elevated levels of cholesterol. Unfortunately, this effect is often accompanied by stomach upset, flushing of the skin, and intense itching. Still more unfortunately, niacin has been promoted to the general public as a do-it-yourself cure for elevated cholesterol; it is not.

CHAPTER 8

Riboflavin

A Vitamin for Depression?

---

*...riboflavin deficiency [proceeds in] the following sequence.
First there is sore throat and the lips,
mouth, and tongue feel as if on fire.
The angles of the mouth become dry and cracked.
The skin of the face, especially the nose, is inflamed.
Finally, the deprived individual may become anemic.*

Riboflavin takes its name from flavus, the Latin word for yellow, and ribose, a simple sugar. Although a variety of water-soluble yellow dyes were studied by biochemists more than a century ago and, in retrospect, signs of riboflavin deficiency had been seen for many years, it was not until the late 1930s that agreement was reached on riboflavin's status as a vitamin. Several reasons for this long delay can be suggested. Riboflavin deficiency does not produce a dramatic and life-threatening disease such as beriberi or pellagra. Diets that are deficient in riboflavin are likely to lack other essential nutrients as well, so that a mixed deficiency disease is seen. Finally, the chemistry of the vitamin B complex turned out to be far more difficult than could be imagined by those who first described "water-soluble B."

The history of riboflavin is inextricably linked with thiamine, the anti-beriberi vitamin, and with niacin, the pellagra-preventative. Indeed, signs now recognized as resulting from

107

*A particularly interesting and persistent idea is that riboflavin is related to depressed mood....
Might it be that normal people neglect their diet, become vitamin deficient, and...enter into a state of depression? Or might it be that mentally disturbed people, especially those who are depressed, lose interest in eating and become vitamin deficient?*

riboflavin deficiency were long regarded as a part of the syndrome of pellagra. Even when riboflavin was isolated in 1933, its role in human nutrition was not clear. Only after niacin was shown to be the pellagra preventative factor could the crucial experiments be undertaken.

The Goldberger pellagra diet consisted of corn meal, cow peas, lard, casein, flour, white bread, calcium carbonate, tomato juice, cod liver oil, and iodide of iron. In 1938, volunteers were fed this diet plus supplements of thiamine, niacin, and vitamin C. As expected, full blown pellagra did not appear. Instead, the lips became red and sore and cracks developed at the angles of the mouth. These signs, formerly associated with pellagra, disappeared completely when riboflavin was added to the diet. Riboflavin was finally ready to join thiamine and niacin in the growing family of B vitamins.

The present view of riboflavin deficiency suggests the following sequence. First there is sore throat, and then the lips, mouth, and tongue feel as if on fire. The angles of the mouth become dry and cracked. The skin of the face, especially the nose, is inflamed. Finally, the deprived individual may become anemic.

A particularly interesting and persistent idea is that riboflavin is related to depressed mood. In the early 1970s, this notion was explicitly tested by the US Army at the Fitzsimons General Hospital in Denver. Six male conscientious objectors were put on a liquid diet for 8 weeks. Riboflavin intake was estimated to be less than 0.1 mg per day (about 5% of the present RDA). Because of the short duration of the study, no classical signs of deficiency were expected or seen. However, a widely used measure of emotional status, the Minnesota Multiphasic Personality Inventory (MMPI), revealed a number of changes concurrent with riboflavin depletion. Since then, several surveys of patients admitted to psychiatric institutions have found that deficiencies of thiamine, riboflavin, and pyridoxine (vitamin B_6) occur in about one-third.

Might it be that normal people neglect their diet, become vitamin deficient, and, as a result, enter into a state of depression? Or might it be that mentally disturbed people, especially those who are depressed, lose interest in eating and become vitamin deficient? Put another way, are we depressed because we don't eat well, or do we not eat well because we are depressed? There are no certain answers to these questions, but it is plausible that the brain is as susceptible to the effects of deranged nutrition as is any other organ of the body.

When the Recommended Dietary Allowances were first issued in 1943, the value for riboflavin was 0.7 mg per thousand calories; for an adult consuming 2700 calories in a day, that would be 1.9 mg. The intervening years have seen the value move up and down over a fairly narrow range. The RDA tables published in 1989 make distinctions on the basis of age, sex, pregnancy, and lactation. At ages 19–50, the values are 1.7 and 1.3 mg for men and women, respectively. The

---

...are we depressed because we don't eat well or do we not eat well because we are depressed? There are no certain answers to these questions, but it is plausible that the brain is as susceptible to the effects of deranged nutrition as is any other organ of the body.

---

maximum value, 1.8 mg per day, is for a woman age 11–22 who is breastfeeding her baby.

Before we consider sources of riboflavin, let's look at the clinical studies that form the basis for the present RDA and for the mild controversy that surrounds those values. The Goldberger pellagra diet provides about 0.5 mg and, as has already been mentioned, causes riboflavin deficiency. During World War II, Ancel Keys at the University of Minnesota maintained six young men at 0.6–0.8 mg for 22 weeks and no effects were detected. In the late 1940s, M. K. Horwitt treated patients at the Elgin (Illinois) State Hospital with five basal diets ranging in riboflavin content from 0.55 to 1.6 mg per day. At 0.8 mg, only one of 22 patients had symptoms. However, intakes of 0.55 mg per day caused "incontrovertible signs of deficiency" involving the mouth and skin. These signs appeared as early as the fourth month and were present in all by the eighth month.

Taken together, the dietary studies conducted between 1938 and 1950 suggest that overt riboflavin deficiency is avoided if the diet provides about 0.6 mg of the vitamin per day. Thus, the present RDA of 1.2–1.8 mg would certainly appear to be ample. However, there are some who disagree. Their arguments are not based upon direct signs of deficiency; these are rarely seen in the United States except in alcoholics. Instead, they rely upon a "biochemical measure of deficiency." This measure depends upon the fact that a decrease in riboflavin intake leads to changes in red blood cells long before effects on the skin are evident. Specifically, the "erythrocyte glutathione reductase activity coefficient" is used as an index of riboflavin status. I'll refer to it as the EGRAC.

Based upon measurements of the EGRAC, a number of surveys have suggested that 10–20% of the US population has a less than adequate intake of riboflavin. Especially prone to deficiency are infants and adolescents living in poverty. The essential feature appears to be a poorly balanced diet that includes little or no milk. The reasons for a low intake

> *...it has been suggested that a variety of "stresses"
> increase the need for riboflavin: pregnancy,
> lactation, trauma, disease, diarrhea, the use of oral
> contraceptives, and so on. None are proven.*

of milk may be both economic and biochemical. Many children, especially blacks and orientals, are unable to digest lactose, the sugar found in milk. The result is that the drinking of milk is followed by considerable discomfort, truly "indigestion," which in turn leads to avoidance of this excellent source of riboflavin.

An unanswered question is whether marginal deficiency of riboflavin, as indicated by the EGRAC, has any effect on health and well-being. Attempts have been made to relate it to stunted growth, cataract formation, aging, cancer, and other hazards, but the best that can be said is that the significance of "biochemical deficiency" is unknown. Likewise, it has been suggested that a variety of "stresses" increase the need for riboflavin: pregnancy, lactation, trauma, disease, diarrhea, the use of oral contraceptives, and so on. None are proven. For example, of a dozen studies that examined the effects of oral contraceptives on riboflavin status, six suggested an increased need for the vitamin and six did not.

One kind of stress that deserves mention, at least in this book, is exercise. In a study reported in April 1983, Amy Belko, Daphne Roe, and their associates at the Division of Nutritional Sciences at Cornell University tested whether jogging increases women's need for riboflavin. The investigators chose a value for the EGRAC that they considered to be "within the normative range" (1.2). They then determined the amount of dietary riboflavin needed to maintain that value in women who jogged 20–50 minutes per day and in women who did not. From their data, the authors concluded that the present RDA is set too low; they suggest that normally active women require about 2 mg per day, whereas those engaged in strenuous exercise need even more, about 2.8 mg.

Anyone of an argumentative nature might choose to challenge the conclusions of the Cornell group on the basis that there is no agreement on what constitutes "biochemical normality" as measured by the EGRAC. The sellers of vitamins are of course not inclined to be argumentative when someone suggests an inadequate RDA. Less than a year after their findings appeared in the *American Journal of Clinical Nutrition*, the Cornell group was mentioned in the 1984 catalog of the General Nutrition Corporation. On page 10 of that catalog, GNC offers for our consideration "Big 150 B Complex" tablets. As the name implies, this "mega star of the B vitamin family" contains 150 mg of riboflavin. If you prefer that in terms of the RDA, it comes out to 8823%.

Daily riboflavin supplements of 10–150 mg or more make no sense whatsoever; even in a maximally depleted state, the human body can absorb no more than about 6 mg in a day. But, what if the Cornell group and others are correct and the RDA is set too low, especially for those who exercise regularly? Can we expect to obtain 3 mg of riboflavin from any reasonable diet or is a modest supplement of perhaps 1.5 mg necessary? We can begin to answer these questions by considering the riboflavin content of various possible components of the diet.

In 1943, a group of patients at the Rochester (Minnesota) State Hospital was fed a diet that excluded all milk, cheese, eggs, organ meats, legumes, and green leafy vegetables for 10 months. Even with this severely restricted regimen, it was estimated that the intake of riboflavin was about 0.7 mg per day. If we now add to this diet a quart of milk (1.7 mg riboflavin) and 3 ounces of chicken (0.7 mg), we would have a total of 3.1 mg. Cut back to one pint of milk, but add 1 ounce of a fortified breakfast cereal and 3 ounces of ground beef on a roll made from enriched flour, and the total riboflavin is 3.3 mg. With these examples before us, it's not hard to imag-

> *...moderate amounts of skim milk, whole grains,*
> *and enriched flour are easily able to provide*
> *an amount of riboflavin modestly in excess*
> *of the present RDA.*

ine a well-balanced diet that provides two or more times the RDA for riboflavin; more than enough to satisfy those who believe, together with Dr. Belko and her associates, that the present RDA is too low.

I've mentioned enriched flour. Because riboflavin is contained in the bran coat of cereal grains, there is significant loss of the vitamin in milling and polishing. For example, unenriched flour has only about one-third the riboflavin of whole wheat. For this reason, there was a Federal requirement during World War II that all flour be enriched with riboflavin, thiamine, niacin, and iron. Most states still have such laws in effect and about 90% of the flour produced in this country has B vitamins and iron added to it. Indeed, because of an early error in calculating the amount of riboflavin to be added, enriched flour has slightly more riboflavin than the original whole grains.

Riboflavin is able to stand the heat of boiling and baking with very little loss. As with the other water-soluble vitamins, however, boiling food in a large amount of water for long periods and discarding the water will deplete riboflavin to a significant degree.

Summing Up

Lack of riboflavin in the diet does not lead to a dramatic or life-threatening disease. This is not to say that any of us should knowingly court riboflavin deficiency; changes in the skin of the mouth and face are sure consequences and we remain uncertain whether more subtle effects may occur. It has been proposed that exercise and other forms of stress may increase the need for riboflavin. Whatever the merits of this suggestion, moderate amounts of skim milk, whole grains, and enriched flour are easily able to provide an amount of riboflavin modestly in excess of the present RDA.

CHAPTER 9

Vitamin B$_6$

Morning Sickness and the Premenstrual Syndrome

...real-life science takes time; more than a decade was spent sorting out just six of the B vitamins.

By 1930, there remained no doubt that "water soluble B" was a complex mixture of several essential nutrients. Thiamine, the anti-beriberi vitamin, was easily distinguished from the rest of the B-complex by its rapid inactivation by heat. Teasing out the rest of the B vitamins would prove much more difficult. Workers around the world used a variety of diets to induce deficiencies in mice, rats, rabbits, pigeons, monkeys, dogs, and various other species, and then attempted to cure their animals with factors that had been isolated by diverse chemical techniques. In contrast with the sanitized straight-line version of science that students usually are taught, real-life science takes time; more than a decade was spent sorting out just six of the B vitamins. As each factor was isolated in chemically pure form, it could be tested in the animal models.

Rats fed diets in which the only B vitamin is the anti-beriberi factor soon develop skin problems. Among others, Paul Gyorgy of Cambridge University used this dermatitis in rats as a model for pellagra. The curative factor in yeast extracts was clearly not riboflavin. That vitamin had been purified in 1933 and could easily be tested. These circumstances led Gyorgy to propose, in a letter written on February 6, 1934, the existence of a distinct "rat pellagra preventative factor" that he designated vitamin B$_6$.

Paul Gyorgy is not remembered as the discoverer of a cure for human pellagra. Alas, the rat does not provide a valid model for the human condition. That became evident when Goldberger's "pellagra-preventative factor" was shown to be niacin and found to be inactive against rat dermititis. But fame, if not fortune, came to Paul Gyorgy nonetheless. His vitamin B$_6$ turned out to be a chemical called pyridoxine that is now recognized as an essential human nutrient.

In 1939, S. A. Harris and Karl Folkers reported the structure and laboratory synthesis of vitamin B$_6$. In fact, what we call vitamin B$_6$ is a collective term for three different chemicals: pyridoxine, pyridoxal, and pyridoxamine. They are present in differing amounts in various foods, but the human body converts them all to the same product anyway. The usual form contained in vitamin pills is pyridoxine. As with the other vitamins, the biochemical functions of vitamin B$_6$ are quite diverse and complicated. Certainly a primary role has to do with the metabolism of amino acids.

With the isolation of vitamin B$_6$, large amounts became available for investigations in both animals and in humans. The first indication that vitamin B$_6$ might be essential in the diets of humans came from studies of patients with pellagra. Dr. Tom Spies and his associates at the University of Cincinnati College of Medicine found in 1939 that some disorders in pellagrans that didn't respond to niacin, thiamin, or riboflavin were cured by vitamin B$_6$.

Despite these suggestive findings, there remained no direct evidence of the essentiality of vitamin B$_6$ for humans. Early in 1948, in an attempt to correct this situation, W. W. Hawkins of the Department of Biochemistry of the University of Saskatchewan put himself on a purified diet containing "all necessary known vitamins with the exception of vitamin B$_6$." The synthetic nature of the diet is made clear by Hawkins comment that "the unpalatability of the mixture made it difficult to supply the needs of a moderately active man." After 55 days on the diet, Hawkins had lost nine pounds of an already spare 129. In addition, there was a modest degree of anemia and his blood pressure fell (to what today we would consider healthier levels). But there were no effects that, in Hawkins words, "could unequivocally be considered as resulting from a lack of vitamin B_6." Toward the end of the 55 days he did note "an unusual degree of depression and mental confusion" and suggested that future investigators might wish to consider mental changes more carefully.

The next attempt to produce vitamin B_6 deficiency in humans was made by Selma Snyderman, MD, and her colleagues at the Department of Pediatrics of the New York University School of Medicine during the winter of 1948–1949. Their subjects were two "mentally defective infants" aged 2 and 8 months, respectively. After being on a diet devoid of vitamin B_6 for 76 days, the younger of the infants suffered convulsions. Four days later vitamin B_6 was given and the convulsions stopped. Early in the experiment, the older baby showed signs of anemia, but remained on the diet for 130 days without the occurrence of convulsions. Both babies had been tiny at the start of the experiment; together they weighed less than 20 pounds. Neither gained appreciable weight during their participation. From their experiment, Snyderman and her associates concluded that vitamin B_6 is essential for normal growth, for red blood cell formation, and for appropriate electrical activity in the brain.

In passing it should be noted that Snyderman's induction of vitamin B_6 deficiency in infants could not ethically be repeated today. Indeed, even at the time the work was done, the investigators seemed a bit ambivalent. In their initial report of the experiment in 1950, it was stated that the deficient diet was "given for therapeutic reasons" (though these reasons were never spelled out). Three years later in a lengthier publication it was stated that "this study was designed to determine if pyridoxine were indeed required by the human and what the signs and symptoms of such a dietary restriction would be in the growing infant." They felt justified in using the infants because they were "severely mentally defective." Furthermore, the babies would be receiving "very specialized nursing and medical care" and would thus experience lesser "risk of contracting various contagious illnesses" than would occur in the usual infants' wards. Such arguments would be unlikely to sway a "human subjects review committee" of the 1990s.

In 1951, the American Medical Association reviewed all of the existing evidence on vitamin B_6. They noted the experiments in animals that suggested that the vitamin was essential for mice, rats, dogs, chickens, pigs, and calves. Regarding the need for B_6 by humans, there wasn't much said. The AMA cited the apparent value of the vitamin in pellagrans who didn't respond completely to the other known B vitamins, Hawkins' attempt at self-deprivation, and Snyderman's experiment with babies. Their conclusion was that there is indeed a human requirement for vitamin B_6. No attempt was made to provide a quantitative estimate of that need.

Without a Recommended Dietary Allowance, mere statement of the belief that vitamin B_6 is essential for humans would have little effect on physicians, nutritionists, or the American public. Few multivitamin tablets of the day even included pyridoxine. More dramatic testimony was needed.

Beginning in the early 1950s, a peculiar convulsive disorder arose among infants in various parts of the United States. Peculiar in the sense that the children seemed not to suffer the ancient disease of epilepsy and no other cause could be found. They were healthy at birth and remained so for several months. The convulsions then began. The babies were treated with the standard anticonvulsant drugs of the day, apparently with good effect.

The solution to the puzzle of the convulsing babies was provided in back-to-back articles in the January 30, 1954 issue of the *Journal of the American Medical Association*. Drs.

Clement J. Malony and A. H. Parmalee of Los Angeles and David B. Coursin, a physician in Lancaster, PA, described 60 babies who had two things in common. All suffered unexplained convulsions and all had been fed a liquid infant formula called "SMA." Beginning in 1951, Wyeth Laboratories of Philadelphia, the manufacturer of SMA, instituted more rigorous sterilization of their product. In their desire to kill bacteria they had unwittingly destroyed virtually all of the natural vitamin B_6 that had been present. The SMA-fed infants thus provided the first convincing demonstration that vitamin B_6 is a dietary essential for humans.

Despite the clear evidence that B_6 is required by babies, progress toward a Recommended Dietary Allowance was slow. There were many reasons for this: absence of a clear-cut deficiency disease, lack of a wholly satisfactory way of assessing the B_6 status of the body, the relation between protein intake and B_6 requirement, and even uncertainties about the content and availability of the vitamin in foods. As a result of these and other problems, it was not until 1963 that the Food and Nutrition Board suggested a tentative value of 1.5–2.0 mg per day. There has been a slight escalation since then to the present recommendation of 2.0 mg for adult males and 1.6 mg for females.

Human experience over the past half century leaves little doubt that overt deficiency of vitamin B_6 is a rare event that is usually associated with general malnutrition. It can be produced in pure form only under extreme conditions such as prevailed in the babies fed exclusively a pyridoxine-free formula. Nonetheless, if RDAs mean anything at all, we must ask the question of whether a well-balanced diet can be expected to provide 2 mg of vitamin B_6 per day.

One of the earliest attempts to estimate the intake of vitamin B_6 was directed by Dr. Grace Goldsmith of Tulane University School of Medicine in the late 1950s. One of her purposes was to find out how much B_6 was contained in diets that by the usual standards were "adequate." These data

---

Let's imagine that a woman about to become
pregnant decides to... increase her vitamin B_6
intake...Although her diet is carefully chosen, the
woman recognizes that it can't provide [enough] B_6
so she takes a 4 mg per day supplement.
Then the wheels begin to turn: If my baby will be
healthier....should I not go all the way and get a
bottle of the super-mega-ultra 500 mg tablets so
attractively displayed in the "nutrition" store?
No! In the several million years of human
conception prior to 1939, no embryo or fetus was
ever exposed to 500 mg per day of pyridoxine.
The woman who decides to take this amount now
is in fact conducting an experiment
with her developing baby as guinea pig.

---

would later be used in setting a Recommended Dietary Allowance. Three sets of meals were put together: adequate (high cost), adequate (low cost), and a "poor" diet patterned after those consumed by a group of poor pregnant women in Tennessee. The poor diet provided only about a thousand calories a day. The estimated B_6 contents of the diets were 2, 2.7, and 0.7 mg, respectively. The greater amount in the low cost adequate diet reflects the more liberal use of beans, whole grains, and tuna, all good sources of the vitamin.

Goldsmith and her colleagues concluded that their estimates "provide no support for a concept of widespread dietary lack" of vitamin B_6. More recent estimates have been less optimistic. In 1985, Diana Polley, Richard Willis, and Karl Folkers (the same Dr. Folkers who synthesized the vitamin in 1939) found that only one of 74 Texas college women met the RDA. A few years earlier, Dr. Karen Schuster and associates of the Department of Food Science and Human Nutrition of the University of Florida examined the diets of poor pregnant women in north central Florida. They estimated a daily intake of vitamin B_6 of only slightly more than 50% of the present RDA for pregnant women (2.2 mg). The Florida group also provided evidence of a relationship between maternal B_6 intake and the condition of the babies at the time of birth. Based on these findings, Schuster and Drs. Lynn Bailey and Charles Mahan recommended in 1984 that pregnant women take in a total of 6–8 mg of B_6 per day. The only reasonable way to achieve that intake is through modest supplementation. (The popular Stewart prenatal vitamins and many house brands intended for pregnant women provide 4 mg of pyridoxine. Other multivitamins are quite varied in content.)

Let's imagine that a woman about to become pregnant decides to take the advice of Dr. Schuster to increase her vitamin B_6 intake to about 6 mg per day. Although her diet is carefully chosen, the woman recognizes that it can't provide that much B_6 so she takes a 4 mg per day supplement. Then the wheels begin to turn: if my baby will be healthier at 2 mg than at 0.2 mg and healthier yet at 6 mg, should I not go all the way and get a bottle of the super-mega-ultra 500 mg tablets so attractively displayed in the "nutrition" store? No! In the several million years of human conception prior to 1939, no embryo or fetus was ever exposed to 500 mg per day of pyridoxine. The woman who decides to take this amount now is in fact conducting an experiment with her developing baby as guinea pig.

A nearly universal experience during the first third of pregnancy is some degree of nausea. Once assured that morning sickness is a sign that all is going well with a pregnancy and that it will soon go away, most women are easily able to contend with the problem. For that small number of women in whom vomiting is so severe as to threaten the health of mother and embryo, many remedies have been tried over the years. In the early 1940s it was the turn of the recently discovered B vitamins. Thiamin, riboflavin, niacin, pantothenic acid, and pyridoxine all had their day, either alone or in combination. A group at the Baylor University College of Medicine reported great success with intravenous administration of B_6.

More than four decades have passed and vitamin B_6 is still widely recommended for morning sickness, whether mild or severe. In 1980 it was estimated that 25% of all pregnant women in the United states took Bendectin, a combination of pyridoxine and an antihistamine. Generally ignored in all of this is the work by H. Close Hesseltine, an obstetrician at the Chicago Lying-in Hospital. A skeptic, he took the trouble to compare B_6 with tablets and injections that contained none of the vitamin. In this way he could separate the influence that a compassionate physician and the simple passage of time have upon morning sickness from the supposed beneficial effects of vitamin B_6. In 1946, Hesseltine published his conclusions: "...pyridoxine is of no more value than scores of other preparations. On a percentage basis, better results were obtained by the (injection) of sterile water and by placebo tablets...pyridoxine is valueless."

There isn't much vitamin B_6 in mother's milk. If you believe in the wisdom of nature, this suggests that only small amounts are needed by nursing infants. On the other hand, just to be safe, the RDA for nursing mothers is 2.1 mg, nearly as much as during pregnancy. In a letter to the *New England Journal of Medicine* in 1979, Leonard Greentree, MD, expressed his opinion that any supplementation with pyridoxine should be avoided. He cited evidence from rodents that large doses of the vitamin decrease the amount of prolactin, the hormone essential for lactation, in the blood. Others quickly pointed out that the amount of pyridoxine found in prenatal vitamins (1–10 mg) is unlikely to be harmful and that suppression of lactation even by very much larger doses (200–600 mg) is uncertain. This matter illustrates once again that our information seldom is as complete as we might wish. For the nursing mother, it suggests that supplements in excess of 10 mg per day are a bad idea, very unlikely to do any good, and possibly able to do harm.

At the risk of erroneously implying that pyridoxine is a "woman's vitamin," I will mention one other uniquely feminine use. In discussing fat and essential fatty acids in Chapter 2, I told you of the possible role of prostaglandins in the "premenstrual syndrome" (PMS). Although PMS was first described in 1931, it remains today a medical condition of uncertain diagnosis and treatment as well as a popular topic in magazines seeking a female readership. Even psychiatrists have expressed interest; a proposal not yet adopted by the American Psychiatric Association would rechristen PMS as "late luteal phase dysphoric disorder." Over the past dozen years or so, some responsible investigators and many women have become convinced that pyridoxine may have a role in therapy. A very careful study by Melanie Williams and her colleagues at the British division of Hoffmann-LaRoche well illustrates both the problems and the promise of pyridoxine.

Williams defined PMS as "one or more premenstrual symptoms, occurring regularly in the cycle and relieved by menstruation." To measure PMS and the influence of pyridoxine upon it, the Williams' group assessed tension, irritability, depression, lack of coordination, violent feelings, headache, breast tenderness, edema and bloating, and acne. A total of 617 women were assigned to pyridoxine and placebo groups. They were unaware of their group assignments during the test period.

After three menstrual cycles, the code was broken and it was found that the frequency of all nine symptoms was diminished in both groups. For example, more than 70% of the women reported a decrease in "violent feelings" whether they were treated with pyridoxine or placebo. In none of the nine categories did the 10 treatment groups differ in their degree of improvement by more than 20%, but in seven of the nine, pyridoxine had a slight edge. When all of the data

---

Anxiety, depression, dementia, autism, fatigue,
nausea, acne, Parkinson's disease: each had their turn.
In none of these conditions, nor in a dozen others one
can dredge up in various vitamin catalogs or "health"
publications, has pyridoxine proven to be a cure.

---

were put together, it was concluded that 82% of the pyridoxine group and 70% of the placebo group were improved.

I draw two conclusions from the investigation we've just considered: (1) PMS is a very tricky condition to study. In this respect it is similar to many other nonlethal, cyclic, and subjectively defined human ills. Unless care is taken to control for investigator and patient expectation, any treatment can be "proven" to be effective. (2) Pyridoxine is a worthy candidate for further effort to define its role in PMS.

Early in this chapter I mentioned that Dr. Tom Spies in 1939 was the first to use synthetic pyridoxine in sick people. The patients were malnourished and clearly suffered from pellagra, but had not responded very well to the usual doses of niacin, thiamine, and riboflavin. Spies listed their symptoms as "extreme nervousness, insomnia, irritability, abdominal pain, weakness, and difficulty in walking." He went on to say that "all patients experienced dramatic relief of these symptoms and increased strength" following the injection of 50 mg of pyridoxine. W. W. Hawkins in his 1948 attempt to induce B_6 deficiency in himself noted "an unusual degree of depression and mental confusion."

In the years following Spies' and Hawkins' reports, each of their notations and many more have been explored. Lou Gehrig's Disease (amyotrophic lateral sclerosis) and other still incurable conditions in which nerves and muscles deteriorate were treated without success with pyridoxine.

Anxiety, depression, dementia, autism, fatigue, nausea, acne, Parkinson's disease: each had their turn. In none of these conditions, nor in a dozen others one can dredge up in various vitamin catalogs or "health" publications, has pyridoxine proven to be a cure.

With all of the possible uses of pyridoxine, one might expect the products of the vitamin peddlers to be loaded with it. However, a reading of current vitamin catalogs reveals a continuing curiosity. To be sure, there are a variety of megadoses of pyridoxine available. The present champion seems to be Puritan Pride's 500 mg tablets; that's 25,000 times the RDA for an adult female. But the same company's "Stress Formula 1,000," "an ultrahigh potency formula to combat the effects of stress," contains only 5 mg. General Nutrition Center's "Super Potent Mega-One" vitamin and mineral supplement contains no B_6 at all! A children's chewable vitamin tablet is said to contain 0.008 mg, a nearly homeopathic dose equal to one-half of one percent of the RDA for a seven to ten year old child. My guess would be that the vitamin sellers assume that the typical buyer of children's vitamins doesn't know enough about RDAs to catch them at this nonsense.

Until very recently it was assumed by all that pyridoxine is completely nontoxic. That changed on August 25, 1983 with the publication in the *New England Journal of Medicine* of the paper "Sensory neuropathy from pyridoxine abuse. A new megavitamin syndrome." The senior author was Dr. Herbert Schaumburg of Albert Einstein School of Medicine in New York City. He was joined in his report by neurologists from the Mayo Clinic and the medical schools of Northwestern University and the University of Pennsylvania.

Dr. Schaumburg and his colleagues described seven adults who had been taking 2000–6000 mg of pyridoxine per day. All had developed difficulty in walking and had begun to lose feeling in their hands and feet. One, a 27-year-old woman, had been told that vitamin B_6 was a "natural" way to rid herself of excess water during a part of her menstrual cycle. Over the course of a year she had increased her daily dose from 500 to 5000 mg a day. By the time she sought medical attention, she could walk only with the help of a cane, had trouble handling small objects, and had severe sensory impairments. After she stopped taking the pyridoxine supplements, her ability to walk slowly improved and she was able to return to work. However, even after seven months she was not completely recovered.

Of the seven people with apparent pyridoxine toxicity described in 1983, none had taken a daily dose less than 2000 mg per day. However, in a subsequent letter Schaumburg and Dr. Alan Berger told of a woman who had received a tentative diagnosis of multiple sclerosis. They found that she had been taking about 500 mg of pyridoxine per day for several years. After she stopped taking the vitamin, there began a slow return to normality. Her case suggests that some individuals may be done harm even by amounts of B_6 previously thought entirely safe.

We've already considered a few of the studies that indicate that many Americans take in much less than the RDA of vitamin B_6. Despite the very widespread distribution of B_6, the content in any one food item is usually small and there probably are significant losses during commercial food processing. As with all the other water-soluble vitamins, prolonged cooking with lots of water can remove much of the vitamin B_6 content. For these reasons it would be easy for me to suggest that everyone take a B_6 supplement. But I won't. A mixed diet rich in whole grains, fruits, and vegetables that includes modest amounts of meat, fish, and poultry will provide adequate amounts of all of the B vitamins, including vitamin B_6. Achievement of such a diet is well worth the effort. To focus upon one or another individual vitamin for supplementation is a foolish distraction. (If you must have a B_6 "supplement," try a banana; an average one contains about 0.6 mg of the vitamin and other good things as well.)

Throughout these chapters I've made no attempt to provide equal time to all the quacks and charlatans who would advise you and me about the use of vitamins, or exercise, or

---

*...it would be easy for me to suggest that
everyone take a B$_6$ supplement. But I won't.
A mixed diet rich in whole grains, fruits, and
vegetables that includes modest amounts of meat,
fish, and poultry will provide adequate amounts
of all of the B vitamins, including vitamin B$_6$.*

---

drugs. In most instances they are easy to ignore; their claims so patently without merit that even an "open-minded" (gullible?) scientist like me quickly dismisses them. Much more difficult to deal with in an honest fashion are those few demonstrably competent scientists who cling to and strongly advocate ideas that are rejected by nearly everyone else. The story of Karl Folkers, pyridoxine, and the carpal tunnel syndrome is an interesting example.

The carpal tunnel syndrome (CTS) is caused by compression of the medial nerve as it crosses between the bones of the wrist and the transverse carpal ligament. The typical CTS patient begins with numbness and strange tingling sensations in the fingers and thumb. This may progress to the point at which the thumb becomes virtually useless, with any movement of the wrist becoming extremely painful. Every textbook of medicine or surgery will tell you that the definitive treatment of CTS is surgical "decompression" of the medial nerve.

Karl Folkers' research career has spanned more than half a century. During his 29 years at Merck and Co. he was responsible for the isolation or synthesis of a variety of vitamins, antibiotics, and hormones. His synthesis of vitamin B$_6$ in 1939 was but one of many accomplishments. At the age of sixty-two he became director of the Institute for Biomedical Research at the University of Texas. In his eightieth year, Dr. Folkers received The Priestley Medal, the American Chemical Society's highest honor. Karl Folkers believes that carpal tunnel syndrome is caused by a deficiency of vitamin B_6. Indeed he believes that 95% of Americans are deficient in B_6 to at least some degree.

In 1976, John Ellis, a physician from Mt. Pleasant, Texas, Folkers, and their colleagues reported the relief of CTS by 100–300 mg per day doses of pyridoxine. Since then they have conducted several more studies, all with the same favorable outcome. Needless to say, orthopedic surgeons did not rush to cancel their CTS surgery and put their patients on pyridoxine. Nor should we expect them to. No treatment can gain acceptance without confirmation by others. Such confirmation has been slow in coming. I am aware of three independent tests of the value of pyridoxine in CTS; the results of only one lent support to Ellis and Folkers.

My point in all this is not to extol the virtues of pyridoxine for CTS. In my mind it remains an unproven treatment. Instead I suggest that the story of Karl Folkers, pyridoxine, and the carpal tunnel syndrome provide a nice example of "alternative medicine" at its best. Pyridoxine, an inexpensive therapy for which some evidence of value has been presented by reputable investigators is tried at nontoxic doses before resorting to surgery, a painful, expensive, hazardous, and less than 100% effective therapy. If pyridoxine fails, nothing has been lost and surgery then can be scheduled.

Summing Up

B_6 is a difficult vitamin to pin down. Aside from convulsions in infants totally deprived of it, no clearcut signs of deficiency are known. It was not until 1963 that an RDA was even specified. Nonetheless, it is worth the effort to insure adequate vitamin B_6 by eating a varied diet. Supplementation with a few milligrams is unlikely to do harm. On the other hand, use of vitamin B_6 in doses hundreds or thousands of times greater than the RDA to treat morning sickness or the premenstrual syndrome is of uncertain value and may cause serious damage to the nervous system.

CHAPTER 10

Pantothenic Acid

The Anti-Gray Hair Vitamin

Gray hair? Balding?
Hair that's thin, limp, and lifeless?
The answer to these problems of life and aging
is that magic member of the B-complex, pantothenic acid.
At least that's what we are told by the advertisements
in the health food stores and in vitamin catalogs.

Gray hair? Balding? Hair that's thin, limp, and lifeless? The answer to these problems of life and aging is that magic member of the B-complex, pantothenic acid. At least that's what we are told by the advertisements in the health food stores and in vitamin catalogs. Eat it, drink it, shampoo with it; we don't care, just buy our pantothenic acid and have thicker, fuller, stronger, more pliable, healthier hair.

The origins of the notion that pantothenic acid is the "anti-gray hair vitamin" are found a half century ago when scientists were struggling to sort out the family of B vitamins. Thiamine, riboflavin, niacin, and pyridoxine (B_6) were generally agreed upon, but there seemed to be other factors as well.

In our discussion of niacin we saw that a disease in dogs called blacktongue was very useful in arriving at the cause and cure of pellagra in humans. Models of pellagra in chickens and in rats were not quite so successful. Indeed, in 1937 what was called "chick pellagra" was found to be something other than a deficiency of niacin. Extracts of liver did cure the disease and scientists set out to identify the "chick anti-dermititis factor." Success was achieved in 1939 with the discovery that the active principal in liver was pantothenic acid, a substance known since 1933 to be essential for the growth of yeast. Pantothenic means "from all sides" and that name was chosen to reflect its widespread distribution in living things.

Chickens identified pantothenic acid for us, but chickens have no hair to turn gray. In September, 1939 the Norwegian scientists Gulbrand Lunde and Hans Kringstad journeyed to the American Chemical Society meeting in Boston to present their studies of nutritional deficiencies in rats. Their provocative title has been a boon to vitamin peddlers ever since: "The Anti Grey Hair Vitamin, A New Factor in the Vitamin B Complex." They proposed to call the factor vitamin B_x. In fact, it had been known for many years that rats fed diets deficient in B vitamins developed diseases of the skin and graying of the fur. With the discovery that chick dermatitis was cured by pantothenic acid, many workers examined the effects of the vitamin in the rat model. The results were a bit puzzling: all agreed that pantothenic acid improved the condition, including a darkening of the gray hair, but some thought that still another factor might be needed. The matter was settled only after general agreement was reached about the existence of biotin, a B-vitamin to be discussed in the next chapter.

As was the case with other vitamins, animals had served humankind well in identifying the nature of pantothenic acid. By the early 1940s all accepted its role in animal nutrition, but nothing was known of human needs. As if on cue, the curtain then rose on World War II. Though it would not be appreciated for several years after the war's end, the period 1941–1945 provided the first clear evidence that pantothenic acid is required by humans, as well as by lower forms of life.

When Hong Kong fell to the Japanese on Christmas Day of 1941, there began a massive study of human malnutrition.

The unwilling subjects were troops of the British garrison. They would soon be joined in captivity by American, Dutch, Canadian, and Australian soldiers in places like Singapore, Java, and the Philippines—wherever the first wave of Japanese victory washed. Fed a generally inadequate diet based on polished rice, it is not surprising that signs of beriberi, pellagra, and riboflavin deficiency began to appear in late spring of 1942. In addition, a peculiar syndrome called "burning feet" began to be seen throughout the prisoner of war camps of the Far East. Nearly 40% of the Americans at Cabanatuan POW camp on Luzon were known to be affected by December, 1942.

John Simpson, a captured medical officer of the British Royal Air Force, described burning feet in the Java camps. "Through the day the men were comparatively free of pain. Nights, however, were spent massaging the feet or incessantly walking in the compound, both acts giving relief. Other men sat with their feet in a bucket of water; they were a pathetic sight during their long nocturnal vigils." Prisoners transported to the Japanese home islands sought relief on the cold brick floors of their barracks. In the bitter winters of Honshu this contributed to frostbite, gangrene, and eventual amputation. With understatement worthy of a British officer, Dr. Douglas Denny-Brown wrote that "confinement in prison camps, on extremely limited diets for long periods of time, has unwittingly provided data on the effect of dietary insufficiencies, on a scale that experimental medicine can hardly hope to emulate."

A few of the captive physicians were aware that burning feet had been reported in various groups of soldiers, laborers, and prisoners as early as 1828 and that a dietary cause had been suggested a hundred years before World War II. In the Japanese prison camps, it soon became evident that the condition was not an aspect of beriberi or pellagra, since neither thiamine nor niacin were curative, nor was riboflavin of any use. It remained for Corish Gopalan, MD, a research assistant in the Stanley Hospital of Coonoor, South India, to provide the answer. Working among the very poor and obviously malnourished people of the area, he demonstrated that pantothenic acid is a specific cure for burning of the feet caused by nutritional deficiency. His results appeared in the January, 1946 issue of the *Indian Medical Gazette*. More than four decades have since passed and Dr. Gopalan's conclusions remain unchallenged.

A diet such as that found in the Japanese prison camps leads to multiple deficiencies and very complex effects. For this reason, scientists much prefer to study subjects in which only one essential nutrient is missing. Beginning in 1951, William Bean and Robert Hodges of the University Hospitals of the State University of Iowa fed volunteers from the Anamosa (Iowa) State Reformatory a variety of diets lacking pantothenic acid. They were unable to demonstrate an unequivocal deficiency syndrome. Only after the diets were combined with a drug that blocks the actions of pantothenic acid did the prisoners become ill. They easily became tired, were irritable, and suffered from vague stomach upsets. More significantly, several of the subjects complained of burning sensations in their feet.

The extent to which the signs and symptoms observed by Drs. Bean and Hodges reflect true pantothenic acid deficiency is uncertain. It is quite possible that the drug used to block pantothenic acid produced other toxic effects as well. In addition, it was found that the prisoners' condition did not immediately improve when pantothenic acid was given in large amounts.

In the absence of a clearly defined and reproducible deficiency disease, it is very difficult to say just how much pantothenic acid humans really need. In 1942, Roger Williams, the discoverer of pantothenic acid, concluded that humans could get along nicely with 11 mg per day. He based this on the pantothenic acid content of what everyone agreed, in 1942, was a good diet: fruits, vegetables, whole grains, and somewhat more meat and dairy products than are now fashionable. A similar approach was taken in the late 1950s by Grace Goldsmith and her colleagues at the schools of medicine of Tulane and Vanderbilt universities. They planned and prepared three sets of meals that they designated "adequate high cost," "adequate low cost," and "poor." The high cost diet provided about 16 mg of pantothenic acid per day, the low cost diet 14 mg, and even the poor diet contained 6 mg per day. The major difference between the diets was in the amount of meat provided; the poor diet had virtually none, whereas meat or fish was included with nearly every meal of the high cost plan. (We should not be misled into thinking that meat is essential to an adequate intake of pantothenic acid. Vegetarians eating a mixed diet get plenty of the vitamin from whole grains, beans, and nuts.) It is quite obvious that finding out how much pantothenic acid is in various diets doesn't tell us how much we need. For this reason, the Food and Nutrition Board of the National Academy of Sciences does not presently offer a Recommended Dietary Allowance for pantothenic acid. Instead, they simply suggest a "safe and adequate" intake. The latest figure is 4–7 mg per day.

T. Clifford Albutt said 80 years ago that "a man of thirty, if he be of liberal education, can read the very newspapers themselves without much harm." Thus reassured I will now mention a few of the largely unproven, and nearly unsupported, claims that present-day vitamin pushers make for pantothenic acid. Rodale Press has published a series of books that are intended to inform the uninformed about the proper way to eat and avoid disease. One of these, *Understanding Vitamins and Minerals*, tells us that pantothenic acid is an "anti-stress vitamin," good for listlessness and fatigue, "remarkably effective in relieving hay fever symptoms," will ease burning feet (you knew that was coming), and may add ten years to your life. Not surprisingly, another arm of the Rodale octopus is happy to sell you pantothenic acid, either alone or in combination with other vitamins and minerals.

Catalogs of [vitamin sellers] are fascinating reading—
filled with strange mixtures of fact, fiction, and fancy;
each copywriter straining to outdo the next
in bringing to us, the uninformed, knowledge
of the latest "nutritional breakthroughs."

Catalogs of companies like Rodale, Puritan's Pride, Nature Food Centres, and General Nutrition Corporation are fascinating reading—filled with strange mixtures of fact, fiction, and fancy; each copywriter straining to outdo the next in bringing to us, the uninformed, knowledge of the latest "nutritional breakthroughs." In a world where words like "ultra," "mega," and "super" are common as fleas on a dog, pantothenic acid receives curious treatment. Puritan's Pride sells a children's multivitamin and a "B-complex" tablet that don't contain any at all. On the other extreme, Western Natural Products offers a tablet containing 500 mg of pantothenic acid. They proudly proclaim this amount to be "5000 percent of the RDA." (Never mind that the Food and Nutrition Board of the National Academy of Sciences has never set an RDA and that 500 mg is closer to 10,000% of the dose range presumed by the FNB to be "safe and effective.") Fortunately for the consumer of super-ultra-mega tablets of pantothenic acid, little harm is likely to result. During the 1950s, volunteers were fed 10–20 grams per day for several weeks without any serious problems arising.

We began this chapter with the tantalizing suggestion that pantothenic acid would prevent, or re-darken, gray hair. For grossly malnourished rats it seems to be true, but over the past 50 years no evidence has appeared that pantothenic acid has anything specific to do with gray hair in humans. Fifty years would seem long enough for an idea without support to die. It hasn't; there are just too many people with

> *In a world where words like "ultra," "mega," and "super" are common as fleas on a dog, pantothenic acid receives curious treatment. [One company] sells a children's multivitamin and a "B-complex" tablet that don't contain any at all. [Another] offers a tablet containing 500 mg of pantothenic acid. They proudly proclaim this amount to be "5000 percent of the RDA."...Fortunately for the consumer of super-ultra-mega tablets of pantothenic acid, little harm is likely to result.*

gray hair who are looking for a simple solution to this universal accompaniment of human aging. Stuart M. Berger, MD, in a book published in 1985 that he modestly titled *Dr. Berger's Immune Power Diet* tells us that "if you are prematurely gray, try this safe, effective formula daily for three months and see if you don't notice a change for the better." The "formula" is 300 mg pantothenic acid, 1000 mg *para*-aminobenzoic acid (a favorite nonvitamin "vitamin" found in health food stores), and 0.6 mg folic acid. Should you or your physician be skeptical, Dr. Berger neatly plants the idea that he knows something that you don't: "Do not be surprised if many physicians are not familiar with immune power nutritional supplements as much of this information is new." (Happily for Dr. Berger and his publisher, but sadly for those without a taste for hogwash, the book eventually reached fourth place on *Time* magazine's bestseller list.)

Even nutritional hypochondriacs (and I include myself in that number) must put pantothenic acid deficiency far down on their list of worries. There is virtually nothing we eat that doesn't contain at least a trace of the vitamin. Whole grain cereals and legumes such as beans and peanuts are

---

...an unsupplemented diet that is adequate
with respect to other B vitamins is guaranteed
to provide enough pantothenic acid.

---

especially rich sources, but substantial amounts are also
found in fruits, vegetables, milk (more than 3 mg per quart),
and meats of all kinds. As always our general principles
apply:

(1) Eat fresh, uncooked, unprocessed fruits and vegetables
    whenever possible.
(2) Don't cook things in lots of water and then throw away
    the water.
(3) Assume that frozen foods retain more of their vita-
    mins than do canned.

Beyond that, rest assured that an unsupplemented diet
that is adequate with respect to other B vitamins is guaranteed to provide enough pantothenic acid.

Summing Up

No clearly defined and reproducible disease is associated with a deficiency of pantothenic acid. However, based
upon experience with grossly malnourished individuals, including prisoners of war, there is little doubt that it is a true
vitamin. At present, no RDA can be given. Instead, a "safe
and adequate intake" of 4–7 mg/day is specified. All but
the most severely unbalanced diets will meet that need. Finally, and with some regret, it must be said that whatever
the effects of pantothenic acid deficiency upon the fur coat
of a rat, the vitamin—even in great excess—will do nothing
for the human variety of gray hair.

CHAPTER 11

Biotin

Did Rocky Know About Avidin?

---

*Stallone's Rocky Balboa staggers to his refrigerator
...cracks five eggs into a glass...drinks the mess down...
[and] starts off on his morning run...
fortified with raw animal energy at its best.
Rocky has never heard of "egg white injury."*

In the predawn Philadelphia darkness, Sylvester Stallone's Rocky Balboa staggers to his refrigerator. He cracks five eggs into a glass and drinks the mess down. The viewer's gag reflex is activated. Rocky starts off on his morning run secure in the belief that he is fortified with raw animal energy at its best. Rocky has never heard of "egg white injury."

In 1922, Margaret Averil Boas was a research fellow at the Lister Institute in London. Her main interests were the roles of calcium and phosphorus in the diet of rats. To minimize those minerals, she fed her rats dried egg white as their sole source of protein. Within three weeks the animals began to lose their hair, their skin became rough and inflamed, and eventually they died. Following up this observation, which others had made before her, Boas conducted a brilliant series of experiments. She found that some foods contain a substance able to protect against injury by egg white; she called it "protective factor X." The material was water soluble and in many respects resembled McCollum's vitamin B. However, closer examination revealed that factor X differed from both the antiberiberi and the pellagra-preventative factors.

Following the publication of Boas' work in 1927, many others took up the search for the identity of factor X. Among them was Paul Gyorgy, whom we met in Chapter 9 during his work with pyridoxine. By 1931, Gyorgy had purified factor X to the point that he was confident that it was in fact a previously unknown vitamin. He proposed the name "vitamin H." (H was simply the next available letter because, two years earlier, the Americans had honored Goldberger by calling the pellagra preventative factor, vitamin G. Neither letter designation is used to any great extent at present.) Animals could be fed raw eggs without harm so long as there was sufficient vitamin H in the diet.

Gyorgy, who by this time had moved from England to the Babies and Childrens Hospital of Cleveland, could get no satisfaction until he knew the chemical identity of his vitamin H. He was aware of the isolation in Germany in 1936 of a factor essential for the growth of yeast; it had been given the name biotin from the Greek word for life. Gyorgy was struck by the similarity of distribution and chemical properties of vitamin H and biotin and by mid-1940 he was ready to conclude that they were in fact identical. Two years later, Vincent DuVigneaud of Cornell University Medical College determined the chemical structure of biotin. (The 1955 Nobel Prize in Chemistry was awarded to DuVigneaud for his synthesis of the hormone, oxytocin.)

The missing piece to the puzzle, the nature of the toxic material in eggs, was provided by a group at the University of Texas. Robert Eakin, William McKinley, and Roger Williams found that chicks fed raw egg white had much less biotin in their tissues despite adequate amounts in the diet. They concluded that there is something in egg white that makes biotin of the diet unavailable for use. In 1941, they showed the material to be a protein for which they suggested the name "avidin." As predicted, avidin is inactivated by heating; cooked eggs do not lead to biotin deficiency.

On October 8, 1941 a retired Italian laborer was admitted to Boston City Hospital. We will know him by his initials, RM (I'd like to think his first name was Rocco.) He had a number of things wrong with him, but the most obvious was the fiery red color of his face and half the rest of his body. RM was attended to by Robert Williams, a junior physician at the hospital. Dr. Williams would later describe RM in the *New England Journal of Medicine*. I quote from that account: "Since adolescence the patient had been extremely fond of raw eggs, putting one or two into each glass of wine that he took...During the six years preceding admission he had drunk from one to four quarts of wine daily. In order to have a sufficient number of eggs for his drinks, he deserted his family and moved to the country so that he could maintain his own chicken farm. During this period of time he ate from two to six dozen raw eggs per week. He did not eat at any regular time. Sometimes he ate only one or two meals a day, and sometimes he drank nothing but wine and eggs for one or two days..."

Although Dr. Williams provided a vivid account of suspected biotin deficiency in humans, credit for the unequivocal demonstration of the syndrome goes to V. P. Sydenstricker. Even before RM was admitted to Boston City Hospital, Dr. Sydenstricker and his associates at the University of Georgia School of Medicine had begun experiments in human volunteers. Their results were described in the February 13, 1942 issue of *Science* and two months later in the *Journal of the American Medical Association*.

Three white men and a black woman were put on a diet low in biotin. Other B vitamins, iron, calcium, and vitamins A and C were given as supplements. To reduce biotin still further, avidin in the form of 200 grams (about 7 oz) of dried egg white was added to the diet. Over the course of 11 weeks on the diet, a variety of signs and symptoms appeared. All of the subjects developed dry and flaky skin. More remark-

---

For anyone able to take food by mouth,
biotin deficiency is not an easy state to attain.
Since 1942, there have been fewer than a half dozen
instances reported in the worldwide medical
literature; all have involved the consumption
of large numbers of raw eggs.

---

able were the mental changes: mild depression at five weeks was followed by sleepiness, lack of energy, and, in two of the subjects, anxiety. The experiment was terminated when all of the patients lost interest in eating. Addition of 150 micrograms of biotin to the diet lead to rapid recovery.

For anyone able to take food by mouth, biotin deficiency is not an easy state to attain. Since 1942, there have been fewer than a half dozen instances reported in the worldwide medical literature; all have involved the consumption of large numbers of raw eggs. Indeed we might be inclined to doubt the reality of the syndrome were it not for the fact that rash and loss of hair has been seen in patients maintained for long periods by intravenous fluids lacking biotin. In addition, a rare disorder of biotin absorption or action was recognized in the late 1970s. Babies born with this "inborn error of metabolism" develop a scaly rash and lose their hair when only a few months of age.

It seems a bit odd that biotin deficiency doesn't develop on even the most bizarre of diets so long as they don't include raw eggs. A possible explanation comes from a curious observation made in the early 1940s: Humans excrete more biotin than they take in. We must recall that human life is not the only life in the human body; we have billions of bacteria living in our guts, and some of these are able to manufacture biotin. The matter is not yet settled, but its seems likely that some of the bacterial biotin is absorbed by our bodies and contributes to overall biotin levels.

All those who sell vitamins are happy to tell you the percentage of the RDA of biotin their product provides. In fact, there is no RDA for biotin. There is so much uncertainty about human needs, the amount contained in a "normal" diet, and the contribution that bacteria of the gut make, that the Food and Nutrition Board of the National Academy of Sciences simply specifies an "estimated safe and adequate level." The range for adults is 30–100 micrograms per day. Even that rather vague estimate is open to question. Everyone agrees that 30–100 micrograms is enough, but many believe that much smaller quantities are needed; perhaps 30–40 micrograms. These lower estimates are based on the fact that typical Western diets that seem adequate in all respects usually contain less than 70 micrograms of biotin.

Occasionally, as in "Nexus Vita-Tress Biotin Shampoo," one finds biotin mentioned in advertisements, but in general not much is made of it. Vitamin World has a 300 microgram tablet, but many multivitamins contain none at all. The same company's "Ultra Vita-Min," a "super high potency, multiple vitamin, mineral, and lipotropic formula," contains 1 (one!) microgram. (I am intrigued by, but know not much about, the corporate identities of the major vitamin peddlers. In the shopping mall that I frequent, Vitamin World abruptly replaced Puritan's Pride. VW's products, claims, and general nutritional hype are identical with PP's. Perhaps Vitamins 'R Us will be next.)

It seems a bit odd that biotin deficiency doesn't develop on even the most bizarre of diets so long as they don't include raw eggs. A possible explanation comes from a curious observation made in the early 1940s: Humans excrete more biotin than they take in.

I suppose I would be reassured in choosing a multivitamin if it contained in the vicinity of 50 micrograms of biotin; at least it would indicate that the company was aware of the vitamin's existence. On the other hand I know of only one instance in which the absence of biotin from a multivitamin made any difference. A misguided physician in Alabama put one of his patients on a diet of six raw eggs and two quarts of skim milk a day for 18 months. That's bad enough, but it happened that her multivitamin was one of those that contain no biotin; the lady entered the medical literature as one of the rare examples of "egg white injury."

Summing Up

There is no doubt that humans need biotin. The precise amount is unknown, but it seems likely that even rather poor diets provide enough. Those who can't resist a supplement can be reasonably sure that no harm will be done; babies with genetically determined biotin deficiency have received as much as 10,000 micrograms per day. My advice is to stay off the raw eggs and let your live-in bacteria worry about biotin.

CHAPTER 12

Vitamin B$_{12}$

Dr. Castle's Predigested Hamburger

---

*Anemia is a condition in which the blood is unable
to carry sufficient oxygen from the air we breath
to the tissues of our bodies....During the 19th
century a number of British physicians
described a form of anemia that so often ended
in death it was given the name "pernicious."*

Anemia is a condition in which the blood is unable to
carry sufficient oxygen from the air we breath to the tissues
of our bodies. The three nutritional factors most commonly
associated with anemia are vitamin B$_{12}$, folic acid, and iron.
In this chapter we will consider the first of these.

During the 19th century a number of British physicians
described a form of anemia that so often ended in death it
was given the name "pernicious." The most famous of those
interested in the disease was Thomas Addison of London
and we still sometimes find reference to "Addisonian" pernicious anemia. In 1824, the Scotsman J. S. Combe suggested
that pernicious anemia is caused by "some disorder of the
digestive...organs." Combe's idea would give birth to a cure
for pernicious anemia; gestation took 100 years.

From the time of its first description there were those
who treated pernicious anemia by dietary means. In general
these efforts met with little long-term success and, in the

absence of any unifying hypothesis of the origin of the disease, there was little incentive to pursue them. Instead, investigators tended to focus on possible defects in bone marrow, the tissue that produces red blood cells, or upon bacterial infection of the gut. Only after scientists became comfortable with the idea of vitamins and the possibility that deficiencies might cause disease was there renewed and sustained interest in the dietary factors of pernicious anemia.

The August 14, 1926 issue of the *Journal of the American Medical Association* carried an article entitled "Treatment of Pernicious Anemia by a Special Diet." The authors were George R. Minot, 40-year-old chief of the medical service at Boston's Huntington Memorial Hospital, and William P. Murphy, only 34, a physician in private practice.

Minot's interest in pernicious anemia began in 1912 when, as a resident physician at the Massachusetts General Hospital, he entertained the thought that the disease might be caused by an inadequate diet. Others before him had regarded diet in rather general terms: The patient with pernicious anemia perhaps needed an especially nutritious diet or an easily digested one. The spark of genius that Minot brought to the question was that food might have some "direct effect on blood." He was encouraged in this line of thinking by experiments done in the early 1920s at the University of Rochester by George Hoyte Whipple and Frieda Robsheit-Robbins. They made dogs anemic by periodic bleeding and showed that specific foods, especially liver, were able to stimulate the formation of hemoglobin, the essential iron-containing protein of red blood cells.

Drs. Minot and Murphy begin: "This paper concerns the treatment of forty-five cases of pernicious anemia in which the patients were given a special form of diet." The most striking feature of the diet was the inclusion of 4 to as much as 10 oz of calf, beef, or lamb liver each day. In general the foods were high in protein and fat and low in carbohydrate; not what we today would call a healthy diet. Not healthy? All

45 of Minot and Murphy's patients quickly improved in appearance and in function. After a month on the diet, their red blood cell counts had more than doubled; in the most profoundly anemic, the increase was nearly fourfold.

We have become accustomed to the calling of press conferences to announce medical "breakthroughs." Today's media-oriented physicians and scientists should blush at the words of Minot and Murphy in 1926:

> It is possible that this series of cases eventually may be proved to be unusual in that there happened to be treated a group that would have taken a turn for the better under other circumstances. Also, time may show that the specific diet used, or liver and similar food, is no more advantageous in the treatment of pernicious anemia than an ordinary nutritious diet. Let this be as it may, at the present time it seems to us, as it has to Gibson and Howard, that it is wise to urge pernicious anemia patients to take a diet of the sort described.

Their modest words were met with enthusiasm in most corners of the world. Though doubts would persist for several decades about the implications of their experiments, most soon accepted the fact that indeed a cure for pernicious anemia had been found. In 1934, the Nobel Prize for Physiology or Medicine was awarded to George R. Minot, William P. Murphy, and George Hoyte Whipple, the first Americans to be so honored.

If pernicious anemia is a disease curable by diet, why don't we all need to eat a half pound or so of liver every day to avoid anemia? William Castle, an assistant resident physician at the Thorndike Memorial Laboratory of the Boston City Hospital, thought he had an answer to that question. Even before the Minot and Murphy diet, Castle was convinced that there was a causal relationship between pernicious anemia and an abnormality in the secretions of the stomach. The way he proved his hypothesis is vividly described in the title of his December 1929 paper: "The effect of the administration to patients with pernicious anemia of the contents of the normal stomach recovered after the ingestion of beef muscle." Dr. Castle provided the "normal stomach."

The experiment began with Castle eating 10 oz of lean ground beef. An hour later he emptied his stomach. The recovered material was incubated to a clear liquid state and then fed by tube to a patient with pernicious anemia. When repeated on a daily basis, this procedure had remarkable effects: The patient looked and felt better and there was an increase in red cells and hemoglobin. Neither normal human gastric juice nor beef digested in the absence of human gastric juice had a beneficial effect. Castle concluded that "some unknown but essential" interaction must occur between beef, the extrinsic factor, and gastric juice, the intrinsic factor.

If Castle was correct, all that remained was to learn the chemical identity of the extrinsic and intrinsic factors. It proved more difficult in the doing than in the saying. By 1937, extrinsic factor had been identified in milk and eggs in addition to beef muscle and liver, but progress in its purification was slow indeed. The major problem was that no one knew how to test for activity of extrinsic factor except in patients with pernicious anemia. They were in short supply and not always willing to be experimented upon. After all, Minot and Murphy had already provided a cure for their disease. As we have come to expect in science, help came from an unlikely place.

Mary Shorb was professor of poultry husbandry at the University of Maryland. She discovered that the growth of a bacterium called *Lactobacillus lactis Dorn* depended upon a factor found in liver. The bacterium could fill in for the hard-to-find pernicious anemia patients.

With Shorb's microbiological assay to guide them, a group of chemists headed by Edward Rickes and including the redoubtable Karl Folkers at the Research Laboratories of Merck and Co. made rapid progress toward isolation of extrinsic factor. In the April 16, 1948 issue of *Science*, they described small red needles of a pure chemical to which they gave the name vitamin B_{12}. In the same volume Randolph West of Columbia University College of Physicians and Surgeons reported three patients with pernicious anemia who responded well to the injection of as little as 3 micrograms of vitamin B_{12}. Vitamin B_{12} behaved in all respects like the anti-pernicious anemia factor of liver. Vitamin B_{12} was the extrinsic factor.

Clinically the problem of pernicious anemia was solved. Vitamin B_{12} could be injected, thus avoiding all problems of absorption from the gut. The fascinating physiological question of the nature of intrinsic factor was answered only slowly, but it turned out that Castle's hunch was correct. We now understand the intrinsic factor to be a protein secreted by specific cells in the stomach. Vitamin B_{12} entering the stomach combines with the protein and the complex is carried to the far end of the small intestine, where B_{12} is absorbed by specific receptors. Victims of pernicious anemia suffer not from a lack of vitamin B_{12} in the diet, but from an inability to secrete the intrinsic factor and a resultant inability to absorb the vitamin.

Before leaving these medical classics, something more needs be said about Minot and Murphy. I've already mentioned that doubts persisted for many years about just what it was that their Nobel-prize-winning diet had done. First of all, Whipple showed in 1936 that his anemic dogs benefited from liver because of the iron it contains. Pernicious anemia patients lack vitamin B_{12}, not iron. Minot's and humanity's good fortune was that the liver-rich diet he chose for the wrong reason happens also to be rich in B_{12}. But liver does not contain intrinsic factor, the B_{12} should not have been absorbed, and so, for a second reason, the Minot-Murphy diet should not have worked.

Forty years after the liver diet was introduced, Ragnar Berlin and his colleagues at the Linkoping University Medical School in Sweden solved the puzzle. They found that about 1% of vitamin B_{12} given by mouth is absorbed even in

---

*All of the vitamin B$_{12}$ required by humans
is produced by bacteria whose humble abodes
are soil, sewage, and the far end of our guts.*

---

the absence of intrinsic factor. Minot and Murphy had cured pernicious anemia because their diet provided B$_{12}$ far in excess of that needed by normal people, but just the right amount for those unable to secrete intrinsic factor.

Now let's turn from the pathological conditions that prevent the absorption of vitamin B$_{12}$ and consider what happens in most of us. Our task is simply to present to our stomachs something on the order of 1 microgram per day of the vitamin.

All of the vitamin B$_{12}$ required by humans is produced by bacteria whose humble abodes are soil, sewage, and the far end of our guts. Though it is true that some members of the animal kingdom eat their own feces (the fancy word for it is coprophagy), there are many good reasons why we humans avoid soil, sewage, and the contents of our large intestines. Instead we allow animals to eat bacteria-laden food, the bacterial B$_{12}$ is incorporated into the tissues of the animal, and we then consume the animal or its renewable parts such as milk or eggs. A neat system for all but those who don't want to eat animal parts; more about them in a moment.

Specification of a Recommended Dietary Allowance for vitamin B$_{12}$ is not an easy matter. Estimates of the minimum amount needed have gone as low as 0.1 microgram per day. The World Health Organization figure, as well as the 1989 RDA, is 2 micrograms per day for adults. During pregnancy and while breastfeeding, the RDAs are increased to 2.2 and 2.6 mcg, respectively. These values are based upon inexact data, such as the amount needed to restore a person with pernicious anemia to health and estimates of the body level and rate of utilization of the vitamin by normal people.

...daily intake of the vitamin is not required....
Even with a total lack of the vitamin in the diet,
a normal adult with maximal stores to begin with
has enough B$_{12}$ *to last for 2–10 years or even longer.*

Although the recommended values for vitamin B_{12}, like those for all other essential nutrients, are on a "per day" basis, daily intake of the vitamin is not required. Substantial amounts are held in reserve and when dietary intake is reduced, the body becomes quite parsimonious; a greater percentage of B_{12} is absorbed and less is excreted. Even with a total lack of the vitamin in the diet, a normal adult with maximal stores to begin with has enough B_{12} to last for 2–10 years or even longer.

Surveys of the American diet indicate that the typical meat-eater has an average daily intake of 5–15 micrograms of vitamin B_{12}. For those who frequently eat organ meats such as liver, the value may be as high as 100 micrograms. As a practical matter, deficiency of vitamin B_{12} in persons with normal gastrointestinal tracts eating a mixed American diet is virtually impossible.

Early in 1977 a 6-1/2 pound baby was born to a 26-year-old woman living in southern California. He was a happy and healthy child for the first 4 months of his life. Then he sickened. He became progressively less active and soon lost the ability to move his head. Feeding at his mother's breast held little interest for him; his weight dropped to 13 pounds. He seemed irritable, but cried little. At 6 months of age he fell into a comatose state, unresponsive even to painful stimuli.

The boy was near death when first seen by Dr. Marilyn Higginbottom and her colleagues at the University of California Medical Center in San Diego. Laboratory tests revealed a profound anemia, together with other evidence of deficiency of vitamin B_{12}. A transfusion of red blood cells pro-

---

Vegetarianism is an emotional subject.
The avoidance of animal products is for many
a matter of religious or moral principle.
This does not mean that the exclusively breast-fed
babies of strict vegetarian mothers need risk
vitamin B_{12} deficiency. Animals are but carriers
of the vitamin; bacteria are the actual source.

---

vided some relief and injections of B_{12} were then begun. Within four days the boy was alert and smiling, he began to gain weight, and his reflexes slowly returned to normal.

The cause of the baby's near-fatal encounter with vitamin B_{12} deficiency was not hard to find. Beginning at age 18, his mother had eaten no animal products—no meat, no fish, no eggs, no milk—and she took no vitamin supplements. Despite this she was not anemic and her B_{12} blood level was in the low normal range, testimony to the body's ability to conserve the vitamin. Nonetheless, her breast milk, the only source of the vitamin for her baby, simply did not contain enough B_{12} for his needs.

Vegetarianism is an emotional subject. The avoidance of animal products is for many a matter of religious or moral principle. This does not mean that the exclusively breast-fed babies of strict vegetarian mothers need risk vitamin B_{12} deficiency. Animals are but carriers of the vitamin; bacteria are the actual source. The "synthetic" form of B_{12} (cyanocobalamin) found in nearly all vitamin supplements is in fact derived from bacterial ferments. Indeed, Dr. Alex Hershaft of the Vegetarian Information Service advises vegetarian mothers to provide their infants and children with B_{12} supplements as "a sensible precaution against possible deficiency."

For those who, in addition to their strict vegetarianism, categorically reject the use of any supplements, the vegetarian community suggests that vitamin B$_{12}$ is provided by certain "sea vegetables" (seaweeds) or by an algae called Spirulina. I can recommend neither source with any degree of confidence because of ongoing uncertainty about their true B$_{12}$ content. But the pregnant woman or nursing mother who places her faith in seaweeds or algae as a source of B$_{12}$ is certainly on firmer ground than she who believes that she can absorb the vitamin from her own intestines.

In their 1985 book *Fit for Life*, Harvey and Marilyn Diamond state that the bacteria residing in our guts produce lots of vitamin B$_{12}$ and that so long as a proper mix of fruits, nuts, and vegetables is provided, no one need worry about deficiency. No doubt exists that human feces contain abundant B$_{12}$, but all agree that it cannot be absorbed from the large intestine. People like the Diamonds, aside from a blind faith in their beliefs, have probably been influenced by a misinterpretation of research done by Drs. V. I. Mathan and S. J. Baker of the Christian Medical College Hospital in Vellore, India.

In 1976 Baker and Mathan found free intrinsic factor in the small intestine of Indian subjects and four years later isolated B$_{12}$-producing bacteria from the same area. They speculated that the increased incidence of B$_{12}$ deficiency among vegetarian Indians after they emigrate from India to England may be caused by a change in the bacteria of their guts despite the maintenance of vegetarian practices. These interesting findings raise the possibility that, under conditions not generally found in Western countries, small amounts of bacterial vitamin B$_{12}$ may be available to the human host. This possibility alters in no way the risk of vitamin B$_{12}$ deficiency in the nursing infant of a strict vegetarian mother.

To this point I have acted as if anemia were the only consequence of deficiency of vitamin B$_{12}$. In fact the brain and spinal cord are directly affected as well. This most commonly results in numbness and tingling in the hands and feet, together with poor coordination. The condition may progress to the point that unaided walking becomes impossible. More

> *Fatigue, headache, constipation, dizziness,*
> *irritability, insomnia, poor memory—*
> *all are possible consequences of deficiency*
> *of vitamin B$_{12}$. In fact...only a vanishingly small*
> *percentage of all the[se afflictions] owes to any*
> *deficiency of vitamin B$_{12}$. The consequence is that*
> *large quantities of the vitamin are used*
> *for conditions totally unresponsive to it.*

difficult to describe and less constant in their appearance are a variety of behavioral effects including changes in mood, poor memory, confusion, and hallucinations. In the past (not in the present, one hopes) these behavioral disturbances have sometimes been confused with mental illness. Numberless poor souls have been sent to mental institutions for lack of a diagnosis of pernicious anemia.

Fatigue, headache, constipation, dizziness, irritability, insomnia, poor memory—all are possible consequences of deficiency of vitamin B$_{12}$. In fact, however, only a vanishingly small percentage of all the fatigue, headache, constipation, dizziness, irritability, insomnia, and poor memory in this country owes to any deficiency of vitamin B$_{12}$. The consequence is that large quantities of the vitamin are used for conditions totally unresponsive to it.

Misuse, some would say abuse, of vitamin B$_{12}$ is not confined to the unlicensed fringe of American medicine. An unknown but undeniably too large number of patients receive regular injections of a lovely red solution of the vitamin. Some of these patients suffer from pernicious anemia; a few have recently undergone surgical removal of large parts of their gastrointestinal tract; for many the injections have no rational basis and the only clear benefit is to the physician's bank account. I prefer to believe that much of the inappro-

---

I am sympathetic to any measure that provides hope
for the afflicted; for most of the history
of medicine, the physician could provide little else.
But worthless treatments that deflect us from
seeking the true causes of illness must be condemned.

---

priate use of B_{12} by physicians arises from ignorance rather than greed, but the consequences for the patient are the same.

A story told by Drs. Larry Lawhorne and David Ringdahl nicely illustrates the benefits ascribed by patients to vitamin B_{12}. Beginning in 1944, the walk-in clinic of a small town in Missouri was visited one-half day per week by the same physician, a general practitioner. Many of his patients became accustomed to receiving B_{12} injections for "fatigue, weakness, and a number of other complaints." Upon the physician's retirement, his practice was taken over by the family practice training program of the University of Missouri School of Medicine. Review of the clinic records showed that 120 patients had been injected with B_{12} on a regular basis during the period 1944–1986.

Further examination of the medical records revealed that only four of the 120 persons receiving B_{12} met accepted criteria for real or potential vitamin B_{12} deficiency. Forty-eight of the patients were still attending the clinic and more information was gathered from them. It was found that one person had been getting the B_{12} shots for 24 years; the average period of vitamin treatment was 10 years. They all felt that the treatment did some good. However, when they were told that "continued therapy was not needed," the majority agreed either to stop the injections and be monitored or to enter a clinical trial. On the other hand, 18 patients, 38% of the group, said that they would seek out another physician willing to continue the injections.

Drs. Lawhorne and Ringdahl conclude that "The efficacy perceived by the patient was more likely related to the support, reassurance, and hands-on care provided by the physician rather than the (B_{12}) injection...." I am sympathetic to any measure that provides hope for the afflicted; for most of the history of medicine, the physician could provide little else. But worthless treatments that deflect us from seeking the true causes of illness must be condemned.

The typical multivitamin tablet contains 1–6 micrograms (usually abbreviated "mcg" on the label) of vitamin B_{12}. Not an unreasonable amount relative to the RDA of 2 micrograms. We must however remind ourselves that none but strict vegetarians need any supplement at all. Upon entering the wonderland of the vitamin hucksters, a several microgram supplement seems an extreme of conservatism. The products sold there provide huge quantities of the vitamin; the current champion is Nature Food Centres' 5000 mcg tablet (that's 166,667% of the RDA). Fortunately, doses of B_{12} even as large as these seem to produce no direct toxicity.

Summing Up

Pernicious anemia was the driving force in the discovery of vitamin B_{12}. Because of the inborn inability to absorb the vitamin from the diet, untreated victims of the disease suffer fatal disorders of the blood and nervous system. For those of us fortunate enough to have a normal digestive system, the risk of deficiency of B_{12} is vanishingly small. An exception are those who eat no animal products; at greatest risk are the exclusively breast-fed infants of strict vegetarian mothers. Vitamin B_{12} has enjoyed a long and undeserved reputation as a "tonic" especially when administered by injection. My inclination is to be wary of anyone, physician, naturopath, or whatever, who recommends B_{12} without clear evidence of need.

CHAPTER 13

Folic Acid

Lucy Wills in the Slums of Bombay

Lucy Wills was trained as a physician in London
[but] gained her place in medical history
[in] the slums of Bombay. She went out to India,
she said, to meet the Viceroy, to see the Himalayas,
and to identify what was killing women in pregnancy
[among whom] a severe and often fatal
form of anemia was not uncommon.

Lucy Wills was trained as a physician in London in the 1920s. She gained her place in medical history not in that city, but in a less elegant part of the Empire, the slums of Bombay. She went out to India, she said, to meet the Viceroy, to see the Himalayas, and to identify what was killing women in pregnancy. Among the poor pregnant women in Bombay, a severe and often fatal form of anemia was not uncommon. The picture of the disease in the blood and bone marrow of its victims was that of Addisonian pernicious anemia. With the discoveries by Minot and Murphy, and by Castle, the matter appeared closed; the anemia was cured by the liver diet and was no doubt caused by a deficiency of Castle's extrinsic factor, the substance we now know as vitamin B_{12}.

Wills was not satisfied; she wanted to know the exact nature of the missing factor. She fed a diet like that of the

poor of Bombay to rhesus monkeys and they became anemic.
When crude liver extracts were added to the diet, the monkeys, like their human counterparts, were cured. But then a
curious thing occurred. As Dr. Wills made the extracts pur-
er, they began to lose their anti-anemic properties. Curiouser yet, the same thing happened in her patients; the purest
liver extracts were inactive. Most curious of all, the purest
liver extracts could still cure Addisonian pernicious anemia.

Dr. Wills' conclusions now have the appearance of in-
evitability, but at the time they were bold indeed.

(1) Bombay's anemia of pregnancy is a disease different
     from that which medicine called pernicious anemia.
(2) Crude liver extracts contain something else than the
     extrinsic factor of Castle. Others would call the some-
     thing else by the name "Wills factor."

In the last chapter we saw how the key to the chemical
identification of vitamin B_{12} was Mary Shorb's discovery that
Lactobacillus lactis Dorn depended upon it for growth. The
chemists were then able to extract and purify and crystallize
to their hearts' delight, guided at each step by the reaction of
the bacteria. In a similar fashion another bacterium would
lead us to the identity of Wills factor.

While Lucy Wills worked in India, scientists in the West
were learning more and more of the peculiar nutritional
needs of other species. Bacteriologists laid claim to the discovery of such odd entities as vitamins B_{10} and B_{11}, norite
eluate factor, *Lactobacillus casei* factor, and factor SLR. From
studies in animals came vitamins M and B_c, and factors 1, U,
R, and S. All of these, together with Wills factor, would
eventually be shown to be one form or another of a chemical
isolated in 1941 by Herschel Mitchell and his colleagues at
the University of Texas. Beginning with four tons of spinach
and *Streptococcus lactis* as their guide, they obtained a substance
they then named folic acid, from the Latin word for leaf, *folium*.
(After its chemical structure was determined, a more proper,

> *New treatments have always had an attraction for*
> *both physician and patient. This is especially true*
> *today, when every minuscule advance is heralded in a*
> *press conference as a "breakthrough." Cynics say that*
> *one should always rush to use a new therapy because*
> *someone will soon show that it doesn't work.*

if less euphonious, name was given—pteroylglutamic acid [PGA]. "Folacin" properly refers to all of the various active forms of folic acid, of which there are more than 100.)

The strikingly similar changes in the blood that occur when we are deficient in either folic acid or vitamin B_{12} suggest a common cause. Today most accept the explanation called the folate trap hypothesis: When too little B_{12} is present in our bodies, folic acid cannot be liberated from its storage sites. The anemia that arises when we are B_{12} deficient is thus identical to that when too little folic acid is present in the diet.

By 1945, the anti-anemia properties of folic acid were well established. Some went so far as to suggest that folic acid was a safe and effective substitute for liver extract in pernicious anemia. This was a serious error. The problem is illustrated by a patient, "OK," treated by Drs. Robert Heinle and Arnold Welch in Cleveland in 1946. A conclusive diagnosis of pernicious anemia had been made and, in keeping with the latest in therapeutic advances, he was treated with 10 mg per day of folic acid.

New treatments have always had an attraction for both physician and patient. This is especially true today, when every minuscule advance is heralded in a press conference as a "breakthrough." Cynics say that one should always rush to use a new therapy because someone will soon show that it doesn't work. Anyway, OK was put on folic acid and within a week all signs of his disease were improved; more red cells were found in his blood and he both looked and felt better. But Mother Nature was playing a not very good joke on OK and his doctors. Three months after starting the folic acid treatment, numbness began to spread from OK's elbows downward; he could neither tie his tie nor button his shirt. When his legs began to lose their feeling, it was obvious to even the most optimistic that a change in therapy was needed. They went back to old fashioned liver extracts and improvement began almost at once.

Just as Wills had suggested a decade before, pernicious anemia is a disease more complex than simple lack of folic acid. Worse, large amounts of folic acid can improve the anemia caused by vitamin B_{12} deficiency while neurological damage goes on. We shall see in a moment how this fact has complicated the question of what constitutes an appropriate folic acid supplement.

Despite the widespread occurrence of folic acid in our foods, there are several situations in which deficiency is likely to occur. The first was illustrated in 1961 by Victor Herbert, then a physician at Boston City Hospital. At that time no one knew what a pure deficiency of folic acid looked like; in a natural setting it is nearly always confounded by multiple deficiencies. Herbert put himself on a diet in which all foods had been boiled three times. By discarding the water after each boiling, most of the water-soluble vitamins were lost. Herbert then added back all but folic acid.

Despite an estimated intake of only 5 micrograms of folic acid per day and a steady decline in the level of folate in his blood, nothing very dramatic happened to Dr. Herbert for more than three months. He then started to have trouble sleeping, he became forgetful and progressively more irritable. Finally, after nearly four and one-half months, the expected anemia appeared. By then taking pure folic acid as a supplement, Herbert concluded that his need for the vitamin was about 50 micrograms per day.

---

*One of the essential functions of folic acid is
to participate in the synthesis of new genetic material.
For this reason, rapidly dividing cells are most
in need of it....Pregnancy is far and away
the most common situation in which humans
experience an increased requirement for folic acid.*

Cultures in which foods are habitually boiled or heated for long periods are in fact replicating the Herbert experiment. It turns out that folic acid is not only water soluble, but heat sensitive as well, so that keeping the cooking water won't help much. But even among people who destroy most of their folic acid in the cooking, a second factor usually must be present before overt deficiency appears. That factor is a greater than normal need for the vitamin.

Pregnancy is far and away the most common situation in which humans experience an increased requirement for folic acid. Magalie, a 24-year-old woman seen by Drs. Mortimer Greenberg and Shirley Driscoll at the Lemuel Shattuck Hospital in Boston illustrates the point. She had never liked fruits and vegetables and, living in a free country, she ate none, apparently without ill effects. Then came pregnancy. Several months after conception, Magalie became severely anemic; laboratory tests revealed a deficiency of folic acid.

One of the essential functions of folic acid is to participate in the synthesis of new genetic material. For this reason, rapidly dividing cells are most in need of it. A baby growing in its mother's womb is a lovely example of a mass of rapidly dividing cells. As a result, lack of enough folic acid is the most common form of malnutrition among pregnant women, especially among poor pregnant women. Magalie's marginal diet was good enough for her alone, but her baby's needs pushed both of them into severe anemia. Even when anemia is not apparent, deficiency of folic acid is associated with miscarriage, excessive bleeding, and generally poorer health of both mother and child. (Other classes of rapidly dividing cells are seen in cancer. Forty years ago drugs were developed with the intention of selectively starving cancer cells of folic acid. That story will be told another time.)

Given the history of folic acid that we have just considered, it is mildly surprising that a Recommended Dietary Allowance was not specified until 1968. The Seventh Edition of the RDAs gave a value of 400 micrograms for adults, 800 micrograms during pregnancy, and 500 micrograms for nursing mothers. Although many believed that these values were unreasonably high, it was not until 1989 that the recommendations were altered. The present RDAs are 180–200 micrograms for adults, 400 mcgs during pregnancy, and 260–280 while breastfeeding. It is true that some vegetarian diets with an emphasis on fresh green vegetables provide much more than 400 micrograms, but typical well-balanced Western diets are more likely to be in the range of 50–300.

The observation that large amounts of folic acid could mask a deficiency of vitamin B_{12} made a deep impression upon leading American physicians and in turn upon those who regulate the availability of vitamins. By the mid-1960s, the Food and Drug Administration ruled that multivitamins containing more than 100 micrograms of folic acid could be obtained only by prescription. As a result, most manufacturers removed folic acid entirely from their products. Of the 46 multivitamin preparations for pregnant women listed in the 1966 Physician's Desk Reference, only five contained any folic acid at all. The stage was set for disaster. Magalie, our pregnant lady in Boston, again illustrates the point. She didn't like fruits and vegetables, but she wasn't stupid. During her pregnancy Magalie took a prenatal vitamin popular in the mid-1960s. Unfortunately it was one of those that contained no folic acid.

A change in attitude toward folic acid supplements has come only slowly and there still is not universal agreement.

---

*A consequence of the controversy surrounding
appropriate folic acid supplements has been a unique
restraint on the part of the vitamin peddlers....
even in the wonderland of the vitamin catalogs,
nothing in excess of 1000 micrograms per pill
is found. Better yet, there seem to be no advocates
of megadoses of folic acid. These facts, together with
an inherent lack of toxicity of the vitamin,
make poisoning with it virtually unknown.*

Most now believe that the hazard to pregnant women of masking anemia caused by a lack of B_{12} with too much folic acid is far outweighed by the risk of anemia caused by too little folic acid. Reflecting this consensus, the FDA has relaxed its control and many multivitamins and prenatal preparations now contain 400 micrograms or more. Nonetheless, there remain on the market a few multivitamins expressly promoted for use by pregnant women that provide no folic acid.

With the change in attitude regarding the relative dangers of folic acid and vitamin B_{12} deficiencies came an interest in the possiblility of fortification of foods. Because of the observation that prolonged heating of foods drastically reduced their folacin activity, it was assumed that fortification of any foods later subjected to cooking would be ineffective. It turns out that although most forms of folic acid found naturally in foods are quite sensitive to heat, folic acid itself is relatively stable. One-third to one-half of folate added to staples such as flour, rice, and corn will survive usual cooking methods. For less developed nations this provides a convenient way to avoid folate deficiency. In the United States, breakfast cereals now quite commonly contain added folic acid and that fact is proudly dispayed on their labels.

A consequence of the controversy surrounding appropriate folic acid supplements has been a unique restraint on the part of the vitamin peddlers. A typical multivitamin contains 400 micrograms, and even in the wonderland of the vitamin catalogs, nothing in excess of 1000 micrograms per pill is found. Better yet, there seem to be no advocates of megadoses of folic acid. These facts, together with an inherent lack of toxicity of the vitamin, make poisoning with it virtually unknown.

Summing Up

As has been the case with every other vitamin we have considered, a mixed diet will provide adequate folic acid without any need for supplementation. On the other hand, sizable segments of the population are at significant risk for folacin deficiency: alcoholics, the poor in general, and the elderly poor in particular, teenagers and others living on burgers and fries and nothing green, and pregnant women. Our most immediate concern must be the last group. After the presence of anemia resulting from vitamin B_{12} deficiency has been ruled out, pregnant women should take a supplement containing a few hundred milligrams of folic acid. There must be continued awareness on the part of both pregnant women and their physicians that some prenatal vitamins continue to provide no folic acid.

CHAPTER 14

Vitamin C

Scurvy, Dr. Pauling, Cancer, and Colds

---

*There are two quite distinct stories that must
be told regarding vitamin C.
The first is concerned with a deficiency disease, scurvy....
The second...has Linus Pauling, twice winner
of the Nobel prize, in the leading role. It is Pauling...
who is responsible for the present level of use,
some would say abuse, of vitamin C in this country.*

There are two quite distinct stories that must be told regarding vitamin C. The first is concerned with a deficiency disease, scurvy, its practical eradication, and the identification of the responsible vitamin. The principal characters are a British sea captain, two Norwegian scientists, and a group of guinea pigs. The second vitamin C story has Linus Pauling, twice winner of the Nobel prize, in the leading role. It is Pauling, more than any other individual, who is responsible for the present level of use, some would say abuse, of vitamin C in this country.

Scurvy is a disease in which small pools of blood appear beneath the skin, the mouth and eyes and skin become dry, the hair falls out, the gums bleed, and there is eventual loss of the teeth. An individual with scurvy is weak and lethargic, everything aches, and there are periods of anxiety and depression. Death may come slowly from infection or suddenly when a weakened blood vessel bursts in the brain.

---

*Scurvy is a disease in which small pools of blood
appear beneath the skin, the mouth and eyes and
skin become dry, the hair falls out, the gums bleed,
and there is eventual loss of the teeth.*

It is thought that scurvy was present in ancient Egypt, in
Rome, and in Greece, during the Crusades, and in many other times and places. But it was with the great voyages of ex-
ploration that scurvy presented itself in full detail. Vasco da
Gama, the first European to reach India by sea, returned to
Portugal with only one-third of his original crew; most of the
rest had succumbed to scurvy. There was hardly an expedi-
tion lasting longer than a few months that was not affected.

Over the years, many remedies for scurvy were suggested. In the early 1500s Jacques Cartier was advised by
the Indians of Canada to use a brew made from spruce needles or sassafras or arbor vitae. Admittedly, many of the
proposals for treating scurvy, particularly those that involved
prolonged heating or fermentation, were almost certainly
without value. However, it is evident that in the 17th century, many men of the sea knew how to cure scurvy. In 1600,
Captain James Lancaster of the East India Company suggested lemon juice and, in books intended for use by ship's
surgeons, the value of fresh fruits and vegetables for treatment of scurvy was acknowledged. In 1617, John Woodall
referred to the juice of the lemon as a precious medicine.

Given this state of knowledge, we may wonder why
sailors continued to suffer and die from scurvy in all navies
of the world. Even the Royal Navy of England was not ex-
empt. For example, Admiral George Anson set out around
the world in 1740 with six ships and a combined crew of 1955.
Four years later he returned with his flagship and 900 men.
Scurvy had gotten most of the rest. The simple answer is
that the minds of naval officers and landlubbers alike were

unprepared for the idea that intimate details of the diet could be responsible for health and disease. We must remind ourselves that the word vitamin did not even enter our language until the early years of the present century.

In the year 1780, the Royal Naval Hospital at Portsmouth cared for 1457 sailors suffering with scurvy; in 1806 there were but two. A single change in the diets of all Royal Navy men was responsible for this remarkable decline. Beginning in 1795, all men at sea for longer than six weeks were to receive one ounce of lemon juice per day. The story of how this change took place is an interesting one for two reasons. First, it is a lovely example of well-designed scientific research and, second, it illustrates the importance of advocacy.

On the 20th of May, 1747, HMS Salisbury was at sea with twelve of her crew sick with scurvy. The ship's surgeon was a 31-year-old Scotsman named James Lind. On that day, Dr. Lind began what Duncan Thomas has called the "first deliberately planned controlled therapeutic trial ever undertaken." He assigned two of the sick men to each of six treatments. These were (1) sea water, (2) vinegar, (3) cider, (4) nutmeg, (5) citrus fruits, and (6) an elixir containing sulfuric acid.

By the end of six days, none but the two receiving the fruit had improved. One of the latter was returned to duty and the other assigned as a nurse to the remaining ten. Lind's demonstration of the efficacy of citrus fruits was to be of little value to the men of the sea for many years. He did not publish his findings until seven years later. (The TV news conference and the *National Enquirer* had not yet been invented.) More important, the Royal Navy officially ignored his demonstration of a cure for scurvy for nearly half a century.

As valuable as was the work by Dr. Lind, credit for the virtual eradication of scurvy from the British Navy must go to James Cook, a man Allan Villiers called "that extraordinary sea genius." Whereas Lind was an obscure surgeon who left the naval service shortly after his tour on Salisbury, Cook was an officer of the line who had commanded vessels in voyages around the world. More important, he had managed to bring back nearly all of those under his command. It is doubtful that Cook thought in terms of any single factor, what we now call a vitamin, keeping his men safe from scurvy, but he surely believed that some aspect of the diet or of general hygiene was of importance. He fed fruits and vegetables of all kinds to every member of his crew. At every opportunity he obtained fresh water and food. He insisted on the highest standards of personal and shipboard cleanliness.

Whatever it was that Cook was doing, it worked; scurvy was eliminated on ships that followed his customs. In 1786, he was elected a fellow of the Royal Society and awarded the prestigious Copley medal for his conquest of scurvy. But, then as now, bureaucracies often had a glacial quality. Not until nine years later did Sir Gilbert Blane, First Lord of the Admiralty, order the use of lemon juice in all ships of the Royal Navy. British seamen, as well as Englishmen at large, became known as limeys because of a misidentification of the Navy's lemon juice as that of the lime. Unfortunately, limes contain only about one-third as much vitamin C, and this had unhappy consequences when, in 1845, scurvy broke out on British ships a few months after an enterprising governor of Bermuda had talked an admiral into substituting Bermudian limes for lemons.

Despite British success in preventing scurvy, nutritional deficiencies continued to diminish the effectiveness of the navies of the world. In the early 1900s, two Norwegian scientists, A. Holst and T. Frolich, wished to develop an animal model for what was called "ship beriberi." It was their great good fortune to choose the guinea pig for study. When they fed polished rice to their animals with the expectation of inducing beriberi, they observed instead a disease "identical in all essentials with human scurvy." Sixty years ago, the authors of a text on vitamins wrote that "the beginnings of our modern knowledge of vitamin C are to be found in the work of Holst and Frolich." Time has not altered that assessment.

Following the discovery that guinea pigs, like humans, are susceptible to scurvy, very rapid progress was made in characterizing the antiscorbutic factor. In some foods, the factor was destroyed by heating. This provided an explanation for what had been a puzzling observation. Infantile scurvy occurred more often in babies fed according to the highest nutritional standards of the times: foods well cooked and milk carefully sterilized.

The existence of an antiscorbutic factor was not immediately accepted by all. Part of the delay arose from the failure to recognize that humans, monkeys, guinea pigs, and a few birds are unique in their dependence on an external source of the factor; rats simply won't do as a model for man. The distinguished American nutritionist, E. V. McCollum, suggested that scurvy in the guinea pig resulted from constipation and that orange juice was merely acting as a laxative (illustrating the fact that when great men err, they often do so in great ways). Nonetheless, by 1919, a prominent British authority felt justified in listing three "accessory food factors": fat-soluble A, water-soluble B, and the antiscorbutic factor, which he called water-soluble C. The case for the existence of vitamin C was essentially closed that same year when an extract assayed for activity in the guinea pig was found to cure scurvy in children. The finishing touches were added between 1928 and 1932, when it was shown that a reducing agent isolated from cabbage and adrenal glands by Albert Szent-Gyorgyi was identical with lemon antiscorbutic factor. When its chemical structure was established, the factor was given the name ascorbic acid. Today, the terms ascorbic acid and vitamin C are often used interchangeably.

What is the human requirement for vitamin C? The first systematic attempt to establish that value was a study conducted in England toward the end of World War II under the direction of Sir Hans Krebs. Nineteen men and one woman lived on a diet that provided less than 1 mg of vitamin C per day. After 17 weeks, signs of scurvy began to

---

*Advertisements for vitamin C stress the need
for daily intake....[such statements] are simply untrue.*

---

appear and became progressively worse. While maintaining
the 1 mg per day diet, the investigators then gave supplements of known amounts of vitamin C. The minimum supplement needed to abolish all of the signs of scurvy was 10
mg per day and subjects remained healthy with that dose for
an additional 14 months. Lesser amounts did not cure the
scurvy; larger amounts seemed to add nothing. The conclusion was that a diet providing 10 mg of vitamin C per day
would prevent scurvy; however, to be safe, the Krebs group
suggested that an intake of 30 mg/day would be appropriate.
A very similar study was undertaken about 20 years later, in
prisoners at the Iowa State Penitentiary by Robert Hodges
and his associates, and the same conclusion was reached.

One may calculate that the prescription of one ounce of
lemon juice by the British Royal Navy in 1795 could have
provided no more than 15 mg of vitamin C and that the
amount would drop to less than 10 mg if lime juice was substituted. Throughout the world, RDAs range from 45 mg in
South Africa and 50 mg in Japan to 75 mg in West Germany
and, highest of all, 100 mg in Russia. In this country, the
RDA was first set at 75 mg in 1943, moved progressively
downward to 45 mg by 1974, and is 60 mg in the 1989
recommendations. A sign of the continuing controversy
surrounding vitamin C is the fact that the committee charged
with revising the Dietary Allowances proposed in 1985 that
the RDA for adult men be cut to 40 mg with comparable decreases for women, infants, and children. Whatever the
merits of that proposal, it was one of the major factors in the
publication delay until 1989 of the Tenth Edition of the RDAs.

Advertisements for vitamin C stress the need for daily
intake. For example, the catalog of Nature Food Centres tells
us that "vitamin C is not stored by the body for future use

(and) that's why you need a fresh supply every day." Statements such as these are simply untrue. A person who consumes the RDA of 60 mg has a considerable amount of the vitamin in storage; normal storage levels are about 1500 mg. Signs of scurvy do not begin to appear until we reach a body level of approximately 300 mg. If we stopped all intake of vitamin C, and nothing else was changed, simple arithmetic indicates that we would not reach that level for 20 days. In fact, we can never make a change in the diet and safely assume that "nothing else was changed." Zero intake promptly results in a decreased rate of metabolism and excretion of the vitamin. If a small amount of the vitamin is still available in the diet, the efficiency of absorption is increased together with the changes in metabolism and excretion. Thus, when attempts have been made to induce scurvy in humans, signs of the disease did not appear for times ranging from 90 to 200 days. The case of Dr. Crandon provides an example.

In 1939, John H. Crandon was an assistant in surgery at Harvard Medical School and a resident at Boston City Hospital. He and his colleagues decided that it would be of value to study the effects of a diet deficient only in vitamin C. They reasoned that all of the earlier accounts of scurvy probably reflected multiple vitamin deficiencies. For a 6-month period, Dr. Crandon ate no fruits and vegetables and drank no milk. During the last four months, his diet was confined to cheese, crackers, bread, eggs, beer, black coffee, and pure chocolate. To prevent other deficiencies, vitamins A, D, thiamin, riboflavin, and niacin, together with yeast tablets and wheat germ oil, were provided. Despite this diet, no signs of vitamin C deficiency appeared until four months had passed. I am not about to suggest that we should go for weeks at a time without ingesting any vitamin C; daily intake is a good idea. However, we should be aware that the body has a sizable store of vitamin C, that fluctuations in day-to-day intake are unlikely to be of consequence, and that missing your orange juice today will not throw you into a scorbutic state tomorrow.

The present RDA of 60 mg includes a large margin for error in that it is six times the amount shown by Krebs and by Hodges to be necessary to prevent scurvy. Nonetheless, there are a few reasonable persons who suggest that the RDA is too low. One of these is Dietrich Hornig, PhD, Director of Clinical Vitaminology for Hoffmann-LaRoche of Basle, Switzerland. Dr. Hornig is not a disinterested observer. Hoffman-LaRoche is the largest manufacturer in the world of ascorbic acid and the company stands to profit by overestimates of the human need for vitamin C. Indeed, it is for that very reason that I find Dr. Hornig's estimates to be of value. In 1981, after a careful review of the existing data, Dr. Hornig concluded that 100 mg per day is adequate "to cover at least 95% of the non-smoking male population." If you should decide that 100 mg is better for you than 60 mg, keep in mind that the difference is equivalent to about 3 oz of orange juice.

Dr. Hornig's specification of nonsmoking males in arriving at his RDA reminds us that certain factors appear to alter one's need for vitamin C. The data are far from complete, but they suggest that blood levels are lower in those who smoke and in users of oral contraceptives. (Dr. Hornig recommends 140 mg a day for smokers.) Gender appears to be a factor. In a recent study conducted in a well-nourished elderly population, it was found that men needed about twice as much vitamin C (150 mg) as did women (75 mg) to maintain the same (rather high) plasma level of the vitamin. Finally, a report from South Africa in 1975 indicated that black miners in that country needed significant supplements of vitamin C in order to maintain what were considered to be adequate blood levels. The authors attributed this to the stress of the workers' occupation, but the study was so poorly designed and controlled that I am unwilling to accept that conclusion.

Not surprisingly, much is made of smoking and stress and oral contraceptives by those who make their living by selling vitamin pills. Full-page advertisements in national magazines have told us that if we smoke, use oral contracep-

> *...much is made of smoking and stress...*
> *by those who make their living by selling vitamin pills....*
> *This attempt to stimulate anxiety about the adequacy*
> *of our diet...[with excessive quantities of vitamin C]...*
> *is more likely to enrich your urine*
> *and the shareholders of [the vitamin sellers]*
> *than it is to alter your response to stress.*

tives, or suffer a stressful life, we had better buy brand x, y, or z of vitamin supplement. For example, Lederle Laboratories, in bold letters, tells us that "Stress can rob you of vitamins" and they define stress as "severe injury or infection, physical overwork, too many martini lunches, fad dieting." The product they are promoting is called, appropriately enough, "Stresstabs 600 High Potency Stress Formula Vitamins" and provides, among other things, 600 mg of vitamin C. This attempt to stimulate anxiety about the adequacy of our diet represents an advertising technique that is unchanged since its introduction in the 1920s by John B. Watson, a psychologist turned advertising man. In any event, regular use of 600 mg of vitamin C is more likely to enrich your urine and the shareholders of Lederle Laboratories than it is to alter your response to stress.

Some establishment nutritionists and physicians become apoplectic at the mention of vitamin supplements. They insist that a well-balanced diet is more than adequate. With that I don't disagree. However, if your diet resembles that of Dr. Crandon (no fruits or vegetables, lots of beer and bread, etc.), a multivitamin containing the RDA of ascorbic acid is a good idea. On the other hand, if you prefer, as I do, to meet all nutritional needs with diet, you will have no trouble at all in obtaining 60 or 120 or even 200 mg of vitamin C each day in ordinary foods. If you like orange juice, you need look no more. Fresh or frozen, it contains about 15 mg per ounce; four ounces will provide the RDA without any help from other sources. In fact, other sources are hard to avoid. For example, the dry cereal that I eat nearly every morning is fortified with 15 mg per ounce. If you happen to be into what I would call weird vegetables, you can forget about scurvy. Things like broccoli, brussels sprouts, kale, and turnip greens contain as much as 30 mg per ounce. There's even one that I like, sweet peppers, in that category. A wide variety of other fruits and vegetables is also a good source; even a baked potato provides about 30 mg. Despite the fact that many of us are remarkably unwise in our choice of foods, the signs of scurvy are rarely seen in the United States except in alcoholics and in grossly malnourished infants and old people. With but a little care, all of us can easily obtain the RDA or several times the RDA in our foods.

And now we arrive at the second story of vitamin C: Linus Pauling, Irwin Stone, common colds, cancer, etc. About 1935, reports began to appear concerning vitamin C and immune function. Vitamin C was used to treat infections such as pneumonia, whooping cough, and rheumatic fever. However, a review of the subject that appeared in the *Journal of the American Medical Association* in 1938 could find no value for it in these diseases. When Crandon reported how he had induced scurvy in himself in 1939, he mentioned "a widespread belief that vitamin C deficiency is an important cause of lowered resistance to infection." However, in the case of the scorbutic Dr. Crandon, "there was an almost complete freedom from respiratory infection throughout the experimental period, covering the months October to May....In previous winters it has not been uncommon for the subject to suffer from frequent, severe upper respiratory infections." Not surprisingly, Dr. Crandon concluded from his personal experience that vitamin C had nothing to do with resistance to colds.

When you or I are coming down with a cold, there is a fall in the level of vitamin C in our blood. It is this observation, more than any other, that has led to speculation that we might be better off if this decrease in vitamin C did not occur. It is reasonable to assume that the fall in vitamim C is caused by its expenditure in fighting the cold and, if only enough were available, the cold would easily be defeated; perhaps we wouldn't even be aware of it. This is an attractive idea that we shall shortly examine in detail. First though, let's consider fever, another aspect of infection. Most of us regard fever as a sure sign of illness and several generations of physicians and mothers have struggled against it, usually by giving aspirin at its first appearance. Only in recent years have we come to suspect that fever may be a defense mechanism that produces an inhospitable environment for the infectious agent. In any event, fever seldom does harm and, save in exceptional circumstances, is best left alone. Might it be that the fall in vitamin C with the onset of infection serves some useful purpose? I don't know the answer, but as G. B. S. Haldane put it, "Nature is not only queerer than we suppose, she is queerer than we can ever suppose."

In 1949, Geoffrey Bourne was a physician at the London Hospital School of Medicine with an uncommon interest in monkeys. Later he became director of the Yerkes Regional Primate Center in the US and one of the world's most distinguished primatologists. In a paper entitled "Vitamin C and Immunity," he began with the question of whether the vitamin C level has something to do with susceptibility to or severity of infection and concluded that there was no satisfactory answer. He then mentioned that great apes in the wild consume 20 pounds or so a day of green feed and thus take in 4–5 grams of vitamin C. "Perhaps," Bourne next wrote, "we should be arguing whether one or two grams per day is the correct amount....We may find that continuous doses of vitamin C at this level over a considerable period of time may have a pronounced and unequivocal anti-infective action."

One of those to hear Bourne's call was Irwin Stone (not to be confused with Irving Stone, author of the popular bio-

---

Linus Pauling...is a man to be reckoned with.
Winner of the Nobel prize for chemistry in 1954
and for peace in 1962, Pauling is both
a distinguished scientist and a world-shaking activist.
In a book published in 1970 called Vitamin C and the
Common Cold, *Pauling advocated the use of several*
grams (several thousand milligrams) per day
in order to attain "the best health."

---

graphies of van Gogh, Darwin, Michaelangelo, etc.). In 1966, Stone, an obscure, 59-year-old chemist from Staten Island, proposed that we all suffer from "hypoascorbemia" (a fancy way of saying "low ascorbic acid in the blood"). Furthermore, hypoascorbemia may be "a very important factor in the incidence and morbidity of diseases, of the aging process, and in the extent of the human life span." In a subsequent book, Stone suggested that vitamin C might be useful for treating colds, polio, hepatitis, herpes, bacterial infections, cancer, heart disease, vascular disorders, arthritis, rheumatism, aging, allergies, glaucoma, cataracts, ulcers, kidney and bladder disease, diabetes and hypoglycemia, the effects of poisons, toxins, and smoking, physical stress, wounds, bone fractures, shock, difficulties of pregnancy, and mental disease. Later he proposed treating narcotic addicts with vitamin C and, in 1983, suggested it as a cure for acquired immune deficiency disorder (AIDS). Stone called for a "National Megascorbic Authority" and a new National Health Institute for "Hypoascorbemia and Megascorbic Disease."

The world does not pay much attention to its Irwin Stones; his ideas might have gained currency in the underworld of nutrition, but little more. Linus Pauling, on the other hand, is a man to be reckoned with. Winner of the Nobel prize for chemistry in 1954 and for peace in 1962, Pauling is both a distinguished scientist and a world-shaking activist. In a book published in 1970 called *Vitamin C and the Common Cold,* Pauling advocated the use of several grams (several thousand milligrams) per day in order to attain "the best health." The public response was overwhelming and, for a time, drug stores had only bare shelves where the vitamin C should have been. There was considerable appeal to the notion that a natural substance—said to be totally harmless and able to be purchased without a physician's prescription—would banish the common cold. As a bonus, Pauling suggested that the vitamin might also be good for back problems and atherosclerosis and would produce a general sense of well being.

The medical establishment was not enthusiastic about Pauling's book. The opinion of reviewers ranged from "near quackery" to "a brilliant but speculative venture." Pauling's writings for a professional audience didn't help. Scientists with a sincere interest in vitamin C's effects could only be offended by a Pauling article in the *Proceedings of the National Academy of Sciences.* He made claims for the value of vitamin C in treating wounds, infection, burns, and shock, but provided no references to investigations by himself or others. In support of the nontoxic nature of vitamin C, Pauling cited Irwin Stone as his authority despite the fact that Stone had conducted no acceptable toxicological tests. (Election to the National Academy of Sciences is one of the highest honors in American science. It carries with it the privilege of publishing, without critical review, in the *Proceedings.* This policy of nonrejection of members' offerings was seriously re-

Nearly all reviews of [Pauling's book] by medically trained persons [asked] why a scientist of Pauling's stature had gone to the public with an idea that he knew had no substantial scientific support and would be accepted uncritically by millions...

---

*My personal opinion is that Pauling wished to force
the medical establishment to test adequately
the hypothesis that vitamin C prevents or cures colds.
If that was his goal, he succeeded.*

---

considered a few years later when Pauling submitted an article claiming value for vitamin C in the treatment of cancer.)

Nearly all reviews of *Vitamin C and the Common Cold* by
medically trained persons raised the question of why a scientist of Pauling's stature had gone to the public with an idea
that he knew had no substantial scientific support and that
he knew would be accepted uncritically by millions of people. Critics would suggest that the answer lay in Pauling's
ego and the attention the book gained him. My personal
opinion is that Pauling wished to force the medical establishment to test adequately the hypothesis that vitamin C
prevents or cures colds. If that was his goal, he succeeded.

Many clinical trials of vitamin C as a therapy for colds
have been conducted. A few of these preceded Pauling's
claims and, depending upon interpretation, could be regarded as either positive or negative. None was fully acceptable by current standards of design. However, in January of 1972, a series of three large-scale, well-designed, trials
was begun by Terence Anderson and his colleagues of the
University of Toronto. Because Pauling had specified in his
book that "the regular ingestion of one gram per day leads
to a decreased incidence of colds by about 45% (and)...to a
decrease in total illness of about 60%," Anderson decided to
give 1000 mg per day and an additional 4000 mg per day at
the onset of a cold. The subjects were assigned randomly to
the vitamin C and placebo groups and neither the subjects
nor the observers were aware of who was in what group,
i.e., the trial was "double blind." The trial continued through
the months of January, February, and March. Of 846 who

started, 818 were present at the end, with dropouts about equally divided between the two groups. When the code was broken, the authors, who had begun the study with the expectation that vitamin C would be without value, found an "entirely unexpected" result. The total days of disability were reduced by 30% in those who had taken vitamin C. The vitamin appeared to have no effect on getting a cold, but the illness seemed less severe. This was not a nonspecific consequence of Pauling's predicted increase in a sense of well being because exactly the same proportion of the placebo and vitamin groups said they felt better. (Had this been an "open" trial, i.e., one without a placebo group, we might have concluded that vitamin C does indeed produce an increased sense of well being in at least 20% of those taking it.)

Encouraged by their initial results, Anderson's group enrolled 2349 people in a study conducted for three months beginning in December 1972. This time an attempt was made to evaluate the ability of three different daily doses of vitamin C to prevent colds (prophylaxis), two different doses to cure colds once started (therapy), the combination of prophylaxis and therapy, and two placebo groups. The outcome of the study was equivocal; critics found the evidence unconvincing, the authors thought that some small benefit in terms of severity of illness had resulted. The major problem in interpretation was that the two placebo groups differed in outcome. When vitamin C was compared with one placebo group, it appeared to have done some good, but when compared with the other, healthier, placebo group, no difference could be detected. Such are the difficult complexities of clinical trials.

In their third and final attempt to settle the question of the efficacy of vitamin C against colds, the Toronto group returned in 1974 to a simpler design. Subjects received either placebo or vitamin C as a 500 mg per day supplement plus 1500 mg on the day a cold started and 1000 mg per day on days two through five. There were three groups in all

---

*The differences between the outcome with placebo
and vitamin C are small and reasonable persons
can disagree as to their importance.*

---

because the vitamin was given as either tablets or sustained release capsules. Any colds that occurred were evaluated in terms of ten different symptoms, so there were 20 comparisons with placebo. There was no change in the number of colds that occurred and only five of the 20 comparisons showed a significant beneficial effect for vitamin C. Critics would again find the study "unconvincing" and there is no question that the extravagant claims made by Pauling were not verified. However, I was impressed by the fact that no one was worse off when taking the vitamin; all 20 comparisons were in a positive direction; we would not expect this to occur by chance. In an admirably conservative assessment of their data, the authors concluded that a small beneficial effect in terms of severity of illness had been demonstrated.

I have provided some of the details of Anderson's studies because his work is as good as or better than any other that has been done in trying to prove Pauling's claims. The differences between the outcome with placebo and vitamin C are small and reasonable persons can disagree as to their importance. However, we are about as close to the truth as we are likely to get for a long time. Effort and money cannot endlessly be poured down the drain of testing speculative and extravagant claims. Indeed, persons less responsible than Pauling are well aware of this fact. They know that any benefit they ascribe to a particular nutrient or diet will (1) be accepted by some people as fact and (2) not rigorously be tested for its truth until long after the money is in the bank.

So, it's time to fish or cut bait: What are we to do about vitamin C and the common cold? My suggestion is that one maintain a daily intake of about 120 mg per day (twice the present RDA) via the diet and that an additional 1000 mg be taken as a supplement on the day a cold starts and for four or five days thereafter. This regimen will provide all that can reasonably be expected from vitamin C. If, at the same time, you throw out all of the cough and cold remedies in your medicine cabinet and swear never to buy them again, you will come out ahead financially. (I would make an exception for pure aspirin or acetaminophen, and perhaps a little hard candy for a scratchy throat.) In 1989, Americans spent in excess of 1.2 billion dollars in treating the common cold with a variety of irrational combinations of drugs that are unlikely to do any good and may cause harm. Vitamin C in moderation, even if it were to act only as a placebo, and there is good reason to suppose that it does something more than that, is a real bargain by comparison.

I have already listed a variety of other conditions for which vitamin C has been recommended. A few of these suggestions, such as its use to reduce cholesterol levels, have adequately been examined in humans and have been rejected. Most of the others remain in limbo somewhere between ridiculous and plausible-but-unproven. An exception is the use of vitamin C for the treatment of cancer. This is worth our attention because of the nature of the disease and because Linus Pauling is again the principal proponent.

So, it's time to fish or cut bait: What are we to do about vitamin C and the common cold?
My suggestion is [an] intake of about 120 mg per day (twice the present RDA) via the diet and...
an additional 1000 mg...supplement on the day a cold starts and for four or five days thereafter. This regimen will provide all that can reasonably be expected from vitamin C.

...the use of vitamin C for the treatment of cancer...
is worth our attention
because of the nature of the disease and because
Linus Pauling is again the principal proponent.

In 1971, Ewan Cameron, a physician at the Vale of Leven
Hospital in Loch Lomanside, Scotland, began to treat cancer
patients with vitamin C. He reasoned that the vitamin would
strengthen "the intercellular ground substance" of normal
tissue and thus permit it "to resist infiltration of malignant
tumors." Daily doses ranged from 10,000 to 50,000 mg. Some
benefit was claimed for all forms of cancer. Not surprisingly, Cameron's work came to the attention of Linus Pauling.

The US branch of Hoffmann-LaRoche used to publish a
small newspaper called "The Good Drugs Do." It was "specifically prepared for the waiting rooms of American physicians to provide patients with accurate information about
drugs, vitamins, and foods their physicians may prescribe"
(Roche's words). An issue appeared in 1975 that was devoted to the possible uses of vitamin C. In it Dr. Pauling stated
that "when a Scottish physician gave 10,000 mg per day to
terminal cancer patients, they felt better and lived longer. In
a controlled study of one hundred cancer patients by the same
physician, long term survival increased fifty fold."

One year later, Cameron and Pauling moved this information from the "waiting rooms of American physicians" to
the *Proceedings of the National Academy of Sciences* under the
title "Supplemental Ascorbate in the Supportive Treatment
of Cancer, Prolongation of Survival Time in Terminal Cancer." In 1978, a second report was published in the *Proceedings* with the subtitle "Reevaluation of prolongation of survival times in terminal human cancer." I quote from the
latter: "There is little doubt...that treatment with ascorbate...
is of real value in extending the life of patients with advanced

cancer. Moreover...the quality of life of the patients is improved...we continue to believe that the addition of ascorbate to treatment regimens at an earlier stage might well have a much greater effect, increasing the average survival time by several years." Both reports by Cameron and Pauling in the *Proceedings* appear to have been based on the same patients to which Pauling alluded in "The Good Drugs Do."

The initiation of medical research in the United States is presently as much a matter of politics as it is of scientific judgment. This is not necessarily bad so long as the political process accurately reflects the informed will of the American people. It is, after all, our money to spend as we wish. Under ordinary circumstances, Cameron's work in Scotland would have been ignored by the American medical establishment in general and the National Cancer Insitute in particular. His studies were too poorly designed and carried out to be taken seriously. However, Pauling's well-publicized advocacy of Cameron's work led to intense political pressure on the National Institutes of Health (NIH) to conduct an adequate trial. As a result, a contract was entered into by NIH and the Division of Medical Oncology and the Cancer Statistics Unit of the Mayo Clinic in Rochester, Minnesota.

The results of the Mayo Clinic trial were reported in the September 27, 1979 issue of the *New England Journal of Medicine*. The title chosen by Drs. Edward T. Creagan, Charles G. Moertel, and their colleagues tells the story: "Failure of High-Dose Vitamin C (Ascorbic Acid) to Benefit Patients with Advanced Cancer." As one would expect from a group at the Mayo Clinic, the design of the study was excellent. Some 150 patients were randomly assigned to either placebo or 10,000 mg of vitamin C per day. Subsequent analysis would show that the groups were well matched for age, sex, site of cancer, stage of the disease, and previous treatment. Neither the patients nor the investigators were aware who received what. When the code was broken, no differences in adverse effects, improvement of symptoms, or survival were found.

How can we explain the difference between the results of Cameron on the one hand and the Mayo investigators on the other? One interesting possibility is suggested by the outcome of a group of 27 patients at the Mayo who initially elected to enter the study, but later changed their minds and received neither placebo nor vitamin C. As a whole, this group survived for only half as long as did those who took part. The explanation for this observation is uncertain. Perhaps the 27 had a more advanced disease or had given up hope or some combination of these and other factors was operating. The point is that if the 27 nonparticipants had simply been compared with those who had received vitamin C, the vitamin would have appeared to double the survival time when in fact it did nothing at all. If we are to arrive at the truth regarding a particular treatment, we must never assume that hope and expectation are of no importance. Indeed, in the Mayo study more than half of these terminally ill patients reported an improvement in symptoms irrespective of whether they received placebo or vitamin C.

Linus Pauling's explanation for the failure of the Mayo investigators to confirm Cameron's findings was that the patient population was different. In a letter to Drs. Creagan and Moertel, Dr. Pauling stated that of the 100 patients treated in Scotland only four had received prior drug treatment and only 20 had received irradiation. Dr. Pauling expressed his belief that the prior drug treatment "may have negated the benefits of vitamin C" for the Mayo Clinic patients. Creagan and Moertel, good scientists that they are, could only reply that, based on Pauling's statement, their study groups were indeed different.

Dr. Pauling, his faith in vitamin C intact, did just what he had done with vitamin C and the common cold: he "went public." In 1979, the Linus Pauling Institute of Science and Medicine, a private organization founded in 1973 to explore some of Pauling's ideas, published a book by Cameron and Pauling called *Cancer and Vitamin C*; in 1981, *Cancer and*

Vitamin C was reissued in paperback and distributed nationally by Warner Books. In their book, Cameron and Pauling expressed once again their belief that high-dose vitamin C has value "for essentially every cancer patient." I believe it fair to say that scientific and medical authorities in this country and elsewhere were united in their skepticism about vitamin C as a cancer treatment. But the general public is notoriously suspicious of "the medical establishment." Futhermore, only a tiny fraction of those millions who read Cameron and Pauling's book were aware of the details of the Mayo Clinic study: The *New England Journal of Medicine* is not to be found at the supermarket checkout. Pressure upon elected officials and upon physicians responsible for the treatment of cancer patients was intense. Once again, the National Cancer Institute turned to the Mayo Clinic.

The second attempt by the Mayo Clinic investigators to verify the claims made by Cameron and Pauling was described on January 17, 1985 in the *New England Journal of Medicine*. One hundred patients with advanced colorectal cancer received either 10,000 mg per day of vitamin C or placebo. As in their earlier study, strict precautions were taken to insure unbiased assessment of treatment outcome. In addition, in accordance with Dr. Pauling's criticism of the earlier investigation, none of the 100 patients had received any prior treatment with anticancer drugs. The results were unequivocal: "No patient had measurable tumor shrinkage, the malignant disease in patients taking vitamin C progressed just as rapidly as in those taking placebo, and patients lived just as long on sugar pills as on high-dose vitamin C."

While Dr. Pauling was engaged in controversy at the national and international level concerning the value of vitamin C in treating human cancer, things were bubbling at home in Palo Alto as well. At the Linus Pauling Institute of Science and Medicine a study in mice was being conducted by Dr. Arthur Robinson, president, director, and trustee of the Institute. Hairless mice reliably develop skin cancer when

---

I suspect that vitamin C in doses of a few thousand milligrams per day...is nontoxic...none of the large double-blind studies of the common cold found an excess of adverse effects....This is not to say that huge doses are a good idea; only that they aren't likely to kill you, at least in the short run.

---

exposed to ultraviolet light. Robinson examined the effects of diets containing varying amounts of vitamin C on the incidence of cancer and concluded that certain levels of vitamin C did offer some protection. However, he also found that a dose equivalent to 10,000 mg per day in humans increased the incidence of cancer. If accepted as fact, this finding would seriously undermine Pauling's claims for the ability of vitamin C to prevent cancer in humans.

During the summer of 1978, Arthur Robinson was removed from his posts at the Linus Pauling Institute. Robinson, in turn, sued Pauling and the Institute for 25 million dollars. That suit was settled out of court in July of 1983 with Robinson receiving $575,000.

Have we seen the end of vitamin C and cancer? No. Vitamin C is one of thousands of chemicals that will continue to be investigated in animals and in humans so long as any form of cancer remains to be cured. We may hope that all studies will be as meticulously designed, conducted, and interpreted as those at the Mayo Clinic; only in this way can we hope to arrive at the truth. Meanwhile, officials of the Linus Pauling Institute of Science and Medicine will continue to say in their fund-raising letters that "...terminally ill cancer patients receiving vitamin C lived an average of seven times as long as patients not receiving vitamin C therapy."

During the long debate between advocates and critics of the use of large doses of vitamin C, much has been made of possible adverse effects. Advocates say that even enormous doses are completely harmless. Critics have raised the specter of formation of kidney stones, dangerously increased iron absorption, interference with anticoagulants ("blood thinners"), nausea, diarrhea, destruction of vitamin B_{12}, gout, and birth defects. I suspect that vitamin C in doses of a few thousand milligrams per day for a few months is nontoxic. I base this tentative belief on the fact that none of the large double-blind studies of the common cold found an excess of adverse effects among the vitamin groups. This is not to say that huge doses are a good idea; only that they aren't likely to kill you, at least in the short run. On the other hand, if I had a history of gout, or stone formation, or I were a woman about to start a baby growing, I would be inclined to avoid even tiny risks.

The probable explanation for the low toxicity of vitamin C is that the human body actively avoids accumulating large amounts of it. I have already mentioned that when vitamin C intake is reduced, a greater proportion is absorbed from the diet and very little is lost in the urine. These conservative measures are reversed when intake is high; less is absorbed and a large amount is lost in the urine. A person who habitually takes in the RDA of 60 mg per day will have about 0.8 mg of vitamin C per 100 milliliters of plasma (usually referred to as 0.8 mg%). If intake is doubled to 120 mg, plasma concentration will increase to about 1.1 mg% and at 20 times the RDA, we may reach 1.5 or 2.0 mg%. Still greater intake results in little additional change. For example, a person taking a supplement of 7000 mg per day was found to have a plasma level of 2.7 mg%; put another way, ingestion of 100 times the RDA led to only a threefold increase in plasma level. I suspect that this is Nature's way of saying "enough is enough."

Summing Up

Any discussion of the most appropriate daily intake of vitamin C must deal with two issues. The first concerns the

---

*A worthy goal would be to insure that no
American, whether young, old, poor, or alcoholic,
suffers the disease of scurvy; that end will admirably
be served by a mere 30 mg/day of vitamin C.*

---

present RDA of 60 mg and whether it should be adjusted
downward or upward by a dozen milligrams or so. I per-
sonally feel comfortable in suggesting 120 mg because some
may benefit, harm is unlikely to be done, and even that
amount can be easily obtained in most unsupplemented diets. The second issue concerns the credibility of those who
would have all of us consume every day, via vitamin supplements, perhaps a thousand times the RDA. The latter view
has been most strenuously promoted by Linus Pauling and
his followers. As deep as is my respect for Dr. Pauling as a
scientist, I do not find his evidence persuasive with respect
either to the common cold or, more emphatically, with respect to the treatment of cancer. A worthy goal would be to
insure that no American, whether young, old, poor, or alcoholic, suffers the disease of scurvy; that end will admirably
be served by a mere 30 mg/day of vitamin C.

CHAPTER 15

Vitamin D

Hormone or Vitamin?

---

*The disease state that results when bones are
improperly calcified...is called rickets or rachitis;
in adults, osteomalacia.*

Steven Spielberg, Jaws, and numerous aquaria have
made most Americans aware of the sinuous grace of a shark
moving through the water. Much of that ease of movement
comes from the shark's lack of bones. Their skeletons are
made of cartilage, a soft and flexible material, a material so
admirably suited to life in the oceans that sharks have
changed little in several millions of years.

A cartilagenous skeleton would never do for us. Humans live in air, not water; we feel the force of gravity much
more keenly than does the shark. Dense, strong, rigid bones
are needed to support the weight of our bodies. Although
our skeletons begin as cartilage or soft membranes, calcium
and other minerals are soon laid down to form bone. As long
as we have adequate calcium in the diet and are exposed to
modest amounts of sunlight, the formation of bone will
proceed in a perfectly normal fashion. The need for calcium
is obvious; the role of sunlight we soon shall consider.

The disease state that results when bones are improperly calcified has various names. In children it is called rickets or rachitis; in adults, osteomalacia. The consequences of
rickets in growing children are dramatic. The spine and leg
bones bend under the body's weight. In its most severe form
breathing and movement are severely impaired.

> *Although rickets was early described by Hippocrates,*
> *humankind's struggle with the disease really began*
> *only about 300 years ago...By 1850 rickets had*
> *become a major source of disability and early death*
> *among the children of the great cities of...Europe...*

Although rickets was early described by Hippocrates, humankind's struggle with the disease really began only about 300 years ago, and then only in a very limited area of the world. The first detailed description appeared in the mid-17th century. On the Continent it was called the "English Disease" because of its prevalence in that country. By 1850 rickets had become a major source of disability and early death among the children of the great cities of Western Europe and Great Britain. Those who survived it were marked for life by bowed legs and misshapen teeth. Poverty was an inconstant factor; the children of factory workers and the unemployed were often afflicted, yet rickets never occurred in even the poorest of farm children. It was indeed a disease of the city.

Armand Trousseau was in 1860 Physician-in-Chief of the Hotel-Dieu, Paris' best and most famous hospital. Although he was known around the world for his descriptions and analyses of many diseases, he had a particular interest in rickets. His ideas about the disease were based partly in an observation made 30 years earlier by his countryman, Jules Guerin, that puppies raised in darkness soon became rachitic. The other influence on his thinking was what most physicians of the day regarded as an old wives' tale. Dutch lore had it that rickets could be cured with cod liver oil. Putting these pieces together, Trousseau hypothesized that the disease was caused by an inadequate diet coupled with a lack of sunshine.

A new scientific truth does not triumph by convincing its opponents and making them see the light, but rather because a new generation grows up familiar with it. (Max

Planck said that.) Trousseau's idea that rickets might be influenced by diet was simply unacceptable to the medical establishment of his time. His notion that sunlight might also be a factor was even more ludicrous. Never mind that in 1890 Theobald Palm, a British medical missionary, clearly demonstrated a worldwide correlation between lack of sunlight and rickets. Never mind that Sir John Bland-Sutton in 1889 cured rachitic lion cubs at the London Zoological Gardens by feeding them cod liver oil. Diet and sunlight would have their day, but it would come in the 20th century.

In lectures delivered at the Royal College of Surgeons of England in 1918, Edward Mellanby listed a number of hypotheses regarding rickets. Among them was Von Hansemann's "Domestication Theory": rickets was a result of the conditions associated with civilized life. Another proposed the causes to be confinement in too small houses and imperfect parental care. Mellanby preferred the "dietetic hypothesis."

There are several reasons why Mellanby was favorably disposed toward diet as the cause of rickets. He knew of Bland-Sutton's work at the London zoo. He was quite familiar with the suggestions by Frederick Gowland Hopkins and Casimir Funk that lack of "accessory food factors" or "vitamines" was responsible for diseases such as beriberi, scurvy, and rickets. Most important, Mellanby had conducted, by the time of his lectures at the Royal College of Surgeons, experiments in more than 200 puppies. Fed a diet of bread and skim milk, they rapidly became rachitic. Supplementation of the diet with cod liver oil or other animal fats pre-

---

Those who survived [rickets] were marked for life by bowed legs and misshapen teeth....the children of factory workers and the unemployed were often afflicted, yet rickets never occurred in even the poorest of farm children. It was indeed a disease of the city.

---

vented the disease. Surprisingly, rickets was not prevented by providing more calcium in the diet. Mellanby's conclusion: "...the cause of rickets is a diminished intake of an anti-rachitic factor which is either fat-soluble A (vitamin A), or has a somewhat similar distribution to fat-soluble A."

As convincing as was the dietetic hypothesis to Mellanby and his colleagues of the "London School," it was rejected by others. In Glasgow, a city noted for a very high incidence of rickets, investigators since the turn of the century had emphasized nondietetic factors. Coincident with Mellanby's 1918 lectures, Leonard Findlay of Glasgow reported the results of his experiments with puppies. Findlay's conclusion: rickets is caused by a lack of fresh air and exercise. These two conflicting views, the dietetic hypothesis of Mellanby and the "London School" and the environmental hypothesis of Findlay's "Glasgow School," would be reconciled within a few years. Crucial to that reconciliation was a publication, little noticed at the time, by Kurt Huldschinsky in Berlin, entitled "The cure of rickets by means of artificial ultraviolet light."

By 1919 the pieces to the puzzle of rickets and sunlight and diet were on the table for all to see. In the Autumn of that year, Harriette Chick, a 44-year-old British scientist, began to put the pieces together. She led a group from the Accessory Food Factors Committee of the Medical Research Council to Vienna. The World War had left there much human misery and widespread malnutrition. As a member of the scientific staff of the Lister Institute of Preventive Medicine, Chick was well aware of the most recent advances in nutrition and with the concept of accessory food factors or vitamins.

Early in their visit, Chick's group demonstrated the cure of infantile scurvy with extract of orange and of keratomalacia with butterfat. These may seem to us rather trivial demonstrations of the use of vitamins, but we must recognize the climate in which Chick and her colleagues found themselves. Professor Clemons von Pirquet, director of the Kinderklinik at the University of Vienna, though a very gracious host,

---

The dietary hypothesis and the environmental hypothesis were each correct and each incomplete.

---

firmly held the view that rickets was an infectious disease much like tuberculosis. An inadequate diet might increase the probability of its appearance, but diet could in no way be curative once the disease was present.

During the period of September 1919 to April 1922, Chick and her colleagues conducted a series of experiments in the rachitic children of Vienna's Kinderklinik. They found that rickets was cured equally well by cod liver oil, ultraviolet irradiation, and natural sunlight. The results of their experiments swept away the controversies of the preceding 80 years. On the one hand, the crucial difference between British farm children and city children was seen to be not diet, but sunlight. On the other hand, the children of Greenland and other places as sunless as any city slum were protected by something in their diet's fish oils. The dietary hypothesis and the environmental hypothesis were each correct and each incomplete.

If children in naturally sunless parts of the world are protected against rickets by a vitamin in fish oils and if children exposed to sunlight have no need for fish oils, might it be that sunlight causes children to manufacture that same antirachitic substance? In her 99th year, Dame Harriette Chick would recall that this was "an entirely new concept at the time." But what a concept.

For the first time the dual nature of the antirachitic substance was clearly seen. For humans exposed to adequate sunlight, it is a hormone, produced in the skin and distributed throughout the body; no dietary source is needed. In a sunless environment, the antirachitic substance must be provided by the diet; it is then a vitamin.

While Chick's work went on in Vienna, investigators in the United States were not idle. Elmer McCollum, having moved to the School of Hygiene and Public Health of Johns

Hopkins University, had developed a rat model of human rickets. With it he and his associates confirmed the efficacy of ultraviolet light. More important, they convincingly distinguished, as Mellanby was earlier unable to do, between vitamin A and the antirachitic substance. In 1922 McCollum, Nina Simmonds, and Ernestine Becker wrote that "the power of certain fats to initiate the healing of rickets depends on the presence in them of a substance that is distinct from fat-soluble A. These experiments clearly demonstrate the existence of a fourth vitamin whose specific property...is to regulate the metabolism of bones." For this fourth essential nutrient they proposed the name "vitamin D."

More than a half century has passed since vitamin D got its name. In those years much has been learned, yet many questions remain still unanswered. In 1924, Alfred Hess, a New York City pediatrician, and Harry Steenbock at the University of Wisconsin independently demonstrated that irradiation of certain foods caused the appearance of vitamin D activity. It was later found that there is not one but two vitamin Ds in nature. That formed in plants is called vitamin D_2 (calciferol, ergocalciferol). When sunlight falls upon our skin a related substance, vitamin D_3 or cholecalciferol, is formed. Vitamins D_2 and D_3 are each able to facilitate the absorption of calcium from the diet. (One might imagine that plants growing in bright sunlight would be good sources of vitamin D_2. They are not. In fact none of the plants eaten by humans contain any vitamin D_2. The reason is that the precursor of D_2, ergosterol, is found only in yeasts, fungi, and a few forage plants. On the other hand fish such as the cod are able to manufacture vitamin D_3 without the benefit of sunlight.)

Rickets is the disease that directed the attention of Chick, Trousseau, and Mellanby to vitamin D and its role in the absorption of calcium from the diet. Since it is a disease of bone, it is easy to draw the conclusion that the primary role of calcium is the formation of strong bones. It is not. In the hierarchy of bodily function, an adequate supply of calcium for nerves and muscle takes precedence over bones. As for the nervous system, bones are little more than a convenient place to store calcium for future use. The full implications of that fact have become apparent only in the last dozen years.

In the early 1950s it was noted by Scandinavian scientists that there is a lag of several hours between the administration of vitamin D and the body's response to it. This led to a 20-year-long unraveling of more mysteries of vitamin D. Our present understanding is that vitamin D_3 is formed in the skin, slowly enters the blood, and is carried to the liver. In the liver, D_3 is converted to $25\text{-OH-}D_3$, which is then transported to the kidney where it is stored for future use.

The parathyroid glands are tiny organs located in the neck. Their function in regulating the level of calcium in the blood was recognized after their inadvertent removal during thyroid surgery in animals and in humans. In response to a fall in blood calcium, parathathyroid hormone is released and travels to the kidney. There it triggers the conversion of $25\text{-OH-}D_3$ to $1,25\text{-di-OH-}D_3$. The final form of vitamin D then acts to increase the absorption of calcium from the foods we eat. If too little calcium is present in the diet, vitamin D will cause calcium to be removed from bones. In addition, parathyroid hormone acts directly on bone to mobilize calcium. The net result is that the level of calcium in the blood is maintained at an appropriate level.

The fine points of vitamin D metabolism and the regulation of blood calcium are of no interest to most of us. But as so often happens, these esoteric studies have led to the most practical of applications: the cure of human disease. For many years the crippling dissolution of bone that accompanied chronic kidney failure was unexplained and untreatable. It certainly didn't respond to the administration of vitamin D. With the recognition that healthy kidneys are required to activate vitamin D, the origin of the disease and its treatment were obvious. Patients are now routinely treated with $1,25\text{-di-OH-}D_3$.

The Recommended Dietary Allowance from age 6 months to 24 years has also been set at 10 micrograms of cholecalciferol, an amount equal to 400 IU. At age 25, the RDA is reduced to 300 IU, and beyond age 22, to 200 IU. For pregnant women and those nursing babies, the recommendation is 400 IU. It should be kept in mind that these values are for total vitamin D. For the vast majority of us, these amounts are easily provided by sunlight. Indeed, there is good evidence the body essentially ignores dietary vitamin D, except when we are deprived of sunlight for extended periods of time.

The rather long process of discovering the origins and functions of vitamin D and its dual nature as both hormone and vitamin has left a residue of misinformation and misunderstanding. Because many of the early dietary studies found some benefit in various meats, animal fats, eggs, and milk, one still finds these listed as sources of the vitamin. The fact is that with the exception of fish oils, no food as it occurs in nature is a reliable source of vitamin D. For this reason, foods in this and other developed countries have since the 1930s been supplemented with either vitamin D_2 or vitamin D_3. These supplements are nearly always stated in International Units or as a percentage of the Recommended Dietary Allowance. In the United States nearly all milk sold is fortified with 400 IU per quart of vitamin D.

The toxic properties of vitamin D have been a primary factor both in establishing an RDA and in attempts to regulate the direct sale of the vitamin to the public. Beginning in the late 1920s, the prescription of enormous doses of vitamin D for any disease even remotely connected with calcium or the bones was not uncommon. Intoxication with 3–6 million units produced loss of appetite, nausea, vomiting, and general irritability. Some children were treated with still larger doses and died as a result. At autopsy, calcium deposits were found in many of the normally soft tissues of the body. The most likely cause of death is kidney failure.

In the 1980s a common source of vitamin D intoxication is the inappropriate treatment of osteoporosis, the thinning of

The toxic properties of vitamin D have been a primary factor...in...regulat[ing]...direct sale of the vitamin to the public.

bone that often affects postmenopausal women (*see* Chap. 24). Though it may seem reasonable to use the vitamin in this disease, there certainly is no virtue in consuming toxic doses. C. R. Paterson recently described a group of women poisoned in Scotland. They experienced nausea, vomiting, headache, apathy, fatigue, and confusion after having been prescribed vitamin D_2 for a variety of conditions. The daily doses were from 25,000 to 400,000 international units. Nearly all recent reports of intoxication in adults have been in this range.

Some will look at those doses, 600–10,000 times the maximum RDA, and conclude that concern about self-medication is unfounded. After all, the largest unit of vitamin D usually available in vitamin stores is 1000 units. Before complacency sets in, let's consider a disease called idiopathic infantile hypercalcemia.

Beginning in the early 1950s reports began to appear from England and Switzerland of infants with unexplained elevations of blood calcium. In mild form, the children simply seemed unwell. The most severe cases had impaired kidney function, a peculiar elfin appearance, dense mineralization of the bones, mental retardation, and abnormalities of the heart. Although the case against vitamin D was never proven with absolute certainty, many international groups were led to urge that very careful controls be placed on the supplementation of foods. Today we are left with the strong suspicion that some infants and children are especially sensitive to the toxic effects of vitamin D. As we saw in chapter 5 with vitamin A, more vitamin D is very definitely not better.

If vitamin D is a poison, why haven't all the beach bunnies, surfers, and tanning booth devotees been killed off? The answer lies in the fact that the human body closely regulates not only the synthesis of vitamin D_3, but also its conversion

---

*...sunlight provides most of us with independence
from dietary sources of vitamin D...
milk supplemented with the vitamin is a nearly ideal
source...and...large doses are potentially toxic.*

---

to the activated form. One of the more interesting means of
regulation is by tanning of the skin. The same region of the
light spectrum that converts 7-dehydrocholesterol to D_3 is
also responsible for a tan. As the skin darkens, less vitamin
D_3 is formed. Years ago W. Farnsworth Loomis made the interesting suggestion that white skin evolved to maximize the
vitamin D-producing effects of the sun as humans moved
away from their tropical origins.

Just how much exposure to the sun is needed to insure
an adequate level of vitamin D? Because of uncertainties
about human need and because of variable degrees of pig-
mentation of the skin and intensity of sunlight at different
times and different places, no precise answer to the question
can be given. A vague answer would be "not very much."
A minimum estimate for infants has recently been provided
by Dr. Bonny Specker and her colleagues at the University
of Cincinnati Medical Center: 10–30 minutes per week, depending on how much of the body is exposed. What is certainly not true is the idea fostered by those with vitamins to
sell that any cloudy day calls for a supplement of 400–1000
or more International Units of vitamin D.

I have emphasized that sunlight provides most of us with
independence from dietary sources of vitamin D, that milk
supplemented with the vitamin is a nearly ideal source for
those in doubt, and that large doses are potentially toxic
especially for infants and children. These facts should not
blind us to the threat of rickets and osteomalacia. In most
instances of these diseases, there is a combination of circumstances: nearly total lack of exposure to the sun, coupled
with marginal levels of calcium and an absence of vitamin D

in the diet. Let's consider how such a combination might arise in this or another developed country.

In early 1961, a Pakistani family living in Scotland was treated by physicians at the Royal Infirmary in Glasgow. Because milk and other dairy products were not a regular part of their diet, the family members had a relatively low intake of both calcium and supplemental vitamin D. Despite this fact, the father showed no signs of calcium deficiency. In contrast, his wife and two teenage daughters complained of severe leg pains and difficulty in walking. Physical examination and laboratory tests revealed that rickets was active in the girls and osteomalacia in the mother. Treatment with vitamin D restored the calcium levels to normal, but the girls will have some skeletal abnormalities the rest of their lives.

Why was the father spared calcium deficiency? The most likely explanation is exposure to sunlight. The whole family lived in Glasgow, a city not noted for bright shiny days and with a variable level of sun-screening air pollution. But the father worked out of doors in typical European clothing. The mother and daughters wore traditional Pakistani dress, which leaves only the face uncovered. The teenagers seldom went outdoors except to school. The final factor was their dark skin, appropriate in sun-drenched Pakistan but still another barrier to the generation of adequate vitamin D in Glasgow.

Since 1961, health authorities in Scotland, England, and other relatively sunless countries have made considerable progress in protecting dark-skinned African and Asian emigrants from rickets. The most effective means is by education regarding the danger. But scientifically based education on nutritional matters often has competition. It comes from the vitamin catalogs to which I've so often referred; their error is always on the side of too much. It comes from self-proclaimed experts on nutrition, themselves woefully uneducated except in the techniques of playing on human fears and ignorance to effect a sale. Perhaps most difficult to counter are the nutritional practices based in religion. Philadelphia in the 1970s provides an example.

All the evidence...about vitamin D indicates that its truly natural role is as a hormone, synthesized in the skin by the action of sunlight, unneeded in the diet.

Rickets was diagnosed in 24 children seen at the Children's Hospital of Philadelphia from January 1974 to June 1978. Sixteen of the 24 were Black Muslims, American Blacks who had embraced the teachings of Mohammed. Meat, fish, and dairy products were excluded from the diets of most who were not still being exclusivley breast-fed. All the families practiced purdah, an ancient Muslim custom that in its strictest form keeps women indoors except when wearing veils and long robes. In these children, the absence of sun-derived vitamin D, coupled to a nonsupplemented vegetarian diet, caused rickets. Fortunately, Dr. Steven Bachrach and his colleagues at the Children's Hospital were able to convince the parents that rickets could be avoided without violating their religious beliefs.

Summing Up

Today many seek to avoid the synthetic, opting instead for what they think "natural." (That's one of the reasons the vitamin peddlers like to call D the "sunshine vitamin," it sounds so natural.) All the evidence so far gathered about vitamin D indicates that its truly natural role is as a hormone, synthesized in the skin by the action of sunlight, unneeded in the diet. To protect those habitually deprived of sunlight, there is unquestioned wisdom in adding vitamin D to basic foods such as milk. To protect those denied both sunshine and supplemented milk (house-bound elderly intolerant of milk and exclusively breast-fed infants come to mind), it may be wise to add modest amounts of vitamin D to the food. For the vast majority of us, and our children, direct addition of vitamin D in the form of pills is both unnatural and unwise.

CHAPTER 16

Vitamin E

A Vitamin in Search of a Disease

Sixty years [after] the discovery of vitamin E
[it is] as controversial as ever....particularly...
with those who have, or fear that they will get,
heart disease, cancer, circulatory disorders,
or any of the problems of aging.

Sixty years have passed since the discovery of vitamin E. It is today as controversial as ever. At one time or another it has been advocated for the treatment of several dozen different diseases. Presently, vitamin E is particularly popular with those who have, or fear that they will get, heart disease, cancer, circulatory disorders, or any of the problems of aging. Because these categories exclude hardly anyone over the age of forty, it isn't hard to understand the enormous consumption of vitamin E in this country. In 1989, about 10 million pounds of vitamin E were produced for medicinal purposes; its dollar value is second only to that of vitamin C.

Just after World War I when Herbert Evans began his research at the University of California at Berkeley, it was recognized that normal growth of adult rats required appropriate amounts of all four of the vitamins then known: A, B, C, and D. Evans wondered whether nutrient needs might be different during reproduction and, in the early 1920s, he and Katharine Bishop found that rats fed a rancid lard diet failed to maintain normal pregnancies. They also observed that lettuce was "spectacularly successful" in correcting whatever the deficiency might be. The unknown dietary factor,

designated X, was surely not the same as vitamin C because
only the fatty component of the lettuce was effective.

The subsequent finding that wheat was as beneficial as
lettuce led Evans to a flour mill in the town of Vallejo. In his
own words:

> I found there three great streams flowing from the mill-
> ing of the wheat berry; the first constituted the outer cover
> or chaff; the second the endosperm, the white so-called flour;
> and the third, which came in flattened flakes stuck into such
> units by its oil content, the germ. Night had not fallen that
> day, before all these components were fed to carefully pre-
> pared females...Single daily drops of the golden wheat germ
> oil were remedial. That an oil might enrich the embryo's
> dietary needs for vitamins A and D, the only fat-soluble
> vitamins then known, was negated at once when we add-
> ed the well-known source of vitamins A and D, cod liver
> oil, an addition that did not lessen, but increased and made
> invarible our malady.

When it thus became clear that crude materials such as
wheat germ oil contain a substance essential for reproduction in the rat, factor X was assigned the next letter of the
alphabet and redesignated vitamin E. Some years later, vitamin E was purified by Oliver and Gladys Emerson and
given the name tocopherol, from the Greek words tocos,
childbirth, and pherein, to carry. The key to an important
function of vitamin E is found in Evans' remark that cod liver oil worsened vitamin E deficiency in his rats. Fish oils
contain high levels of polyunsaturated fats. When these fats
combine with oxygen, they yield potentially harmful products called free radicals. Vitamin E is able to prevent the
formation of free radicals from fats, but it is consumed in the
process. Any of Evans' rats with marginal stores of vitamin
E were thus pushed into total deficiency by the addition to
their diet of polyunsaturated fats in the form of cod liver oil.

Advertisements for vitamin E imply that we all are on
the verge of deficiency if not already frankly deficient. There

---

*Vitamin E deficiency has never been observed
in an otherwise normal human.*

---

is no evidence to support such claims. In a study begun in October, 1953, M. K. Horwitt and his colleagues at the Elgin (Illinois) State Hospital tried for over six years to induce signs of deficiency of vitamin E. A group of 19 patients was placed on a diet in which the source of fat was lard from which all vitamin E had been removed. Vitamin E intake from the rest of the diet was about four units per day. After more than two years on the diet, vitamin E in the blood plasma had dropped to one-half of its original value, but no signs of deficiency were seen. To lower vitamin E still further, intake of polyunsaturated fat was increased by substituting corn oil (from which the naturally occurring vitamin E had been removed) for lard. Plasma vitamin E levels were again cut in half and again no signs of deficiency appeared despite continuation of the corn oil diet for an additional four years.

In view of the inability of dietary studies such as those by Horwitt to demonstrate a deficiency disease, one may wonder why we regard E as a human vitamin at all. The fact is that the evidence for its essentiality has come entirely from persons who are unable to absorb it from the diet. For example, in cystic fibrosis, levels of vitamin E tend to be very low and there is a shortening of the life of the red blood cells. In addition, recent evidence suggests that malabsorption of vitamin E is associated with deterioration of the nervous system in humans in a fashion very much like that seen in animals. It is important to keep in mind that these effects occur only in the presence of well-established diseases of the digestive system or in persons who have had large sections of the small intestine removed surgically. Vitamin E deficiency has never been observed in an otherwise normal human.

In 1968, when the Food and Nutrition Board of the National Academy of Sciences first set a Recommended Dietary

---

*Dr. Edward Rynearson, [of] the Mayo Clinic, said that
the list of conditions treated with vitamin E is limited
only by one's imagination; it was M. K. Horwitt
who called it "a vitamin in search of a disease."*

Allowance for vitamin E, it chose a value of 25–30 International Units. Soon after, it was realized that a well-balanced American diet would provide far less than the RDA and, by definition, the intake of nearly everyone in the country would be deficient; good news for the sellers of vitamin supplements, but obvious nonsense. Thus, in the 8th edition (1974) of the RDAs, the value was reduced to 12–15 IU per day. The RDA was unchanged in the 1989 recommendations, but is now expressed in terms of "alpha-tocopherol equivalents" (TE). From age 11 on, the values are 10 and 8 TE for males and females, respectively. For pregnant women and nursing mothers, an increase to 10–12 TE is suggested. Even these values are regarded by some as being unrealistic because one's requirement for vitamin E is influenced by the amount of unsaturated fat in the diet. An individual with a low intake of such fat needs no more than a few units of vitamin E per day, while someone consuming large amounts of polyunsaturated fat may require 50–60 units. Because so many Americans have heeded the advice of the Federal Government and have substituted vegetable oil margarines for butter in an attempt to reduce cholesterol levels, the average intake of polyunsaturated fat is much higher than previously. This does not mean that supplementation with vitamin E is now necessary. The vegetable oil margarines provide more than enough vitamin E to balance their content of unsaturated fat.

Turning from vitamin E as an essential nutrient to the use of the vitamin as a drug to prevent or to treat disease, we enter an area of controversy. Dr. Edward Rynearson, professor of medicine at the Mayo Clinic, said that the list of conditions treated with vitamin E is limited only by one's imagination; it was M. K. Horwitt who called it "a vitamin in search of a disease." In various animal species, deficiency of vitamin E produces a bewildering array of pathologies, many of which resemble human diseases. Over the years this has led to flurries of interest in the use of the vitamin in a variety of disease states. As popularity has waned in one area, it has waxed in another.

When Evans observed in 1926 that male rats deprived of vitamin E suffered degeneration of the testicles and sterility, he gave birth to the myth of vitamin E as an antisterility, anti-impotence agent, highly recommended to this day for men with diminished sexual powers. Unfortunately, there is no evidence that vitamin E influences either impotence or sterility in men. A few years after its discovery, H. J. Metzger and W. A. Hogan of Cornell University reported a "queer degenerative process of voluntary muscle" in lambs. Because the condition was corrected by the addition of vitamin E to the diet, it was called "nutritional muscular dystrophy." One can imagine the hope engendered in persons afflicted by one of the several forms of human muscular dystrophy. Once again, however, animals turned out to be poor models for humans. Vitamin E in doses of thousands of units per day produced no beneficial effect in patients with muscular dystrophy. The use of vitamin E in diseases of the heart and circulation has had great staying power. The vitamin is as popular for these conditions as it was 30 years ago, yet no evidence of its value has come to light. I have selected one of the expressions of vascular disease, angina pectoris, for special consideration because it well illustrates the problems we may encounter in getting at the truth in therapeutics.

When a muscle receives insufficient oxygen to meet its needs, discomfort results. When the muscle is the heart, the discomfort is given the name angina pectoris (literally, a strangling of the chest). The most common form of angina occurs in response to exertion or emotional stress. It is often

---

It is not unusual for the original report of a therapeutic
advance to consist of...even a single patient....
It is an inflexible rule of medical research that a new
treatment can be recommended for general use only
after it has been verified by individuals...
totally independent of the original proponents.

---

described by the patient as a pressure or tightness of the chest
and relief is obtained by rest or relaxation. A frequent cause
of angina is a narrowing of the coronary arteries and decreased flow of blood to the heart muscle by atherosclerosis.
In its issue of June 26, 1946, *Time* magazine described "a
startling medical discovery....A treatment for heart disease
which so far has succeeded against all forms of the ailment....It
eliminates anginal pain..." The treatment was vitamin E, as
administered by Dr. Arthur Vogelsang and Drs. Evan and
Wilfred Shute of the Canadian city of London, Ontario. Thus
was brought to the attention of the American public a therapy for angina that has persisted to the present day in the face
of intense controversy. Evidence presented to the medical
profession by the Canadian physicians consisted of a series
of five monthly articles beginning in January of 1947. Brief
descriptions of individual patients (case reports) were presented and the conclusion reached that vitamin E is effective
in angina pectoris, rheumatic heart disease, hypertension, and
a variety of conditions related to atherosclerosis.

It is not unusual for the original report of a therapeutic
advance to consist of a few case reports or even a single patient. However, it is then necessary for the therapy to be
evaluated by others. It is an inflexible rule of medical research that a new treatment can be recommended for general use only after it has been verified by individuals who are
totally independent of the original proponents. Furthermore,

subsequent studies must be conducted according to certain rules of investigation. The exact design of a study will vary with the disease in question. The simplest possible case is a well-defined condition that is invariably fatal and a treatment that is always successful. For example, the initial use of penicillin against streptococcal infections came close to being a simplest possible case. In contrast, a condition such as angina is variable in its course and influenced by factors as diverse as emotional state and time of day. For these reasons, evaluation of a treatment for angina that is less than completely effective can be very difficult.

In the original report, Shute described 84 patients who had been treated with 150–300 units of vitamin E per day. In 52 of these (62%), angina was eliminated or markedly improved and in only four was the vitamin ineffective. Between 1948 and 1950, ten independent attempts were made by other physicians to reproduce the Canadian results. These studies were variable in quality, but their outcome was the same: no improvement of angina by vitamin E. In light of these ten trials without success, would we expect physicians in general to accept the new therapy and apply it to their patients? We would not. Indeed, we might question the wisdom and competence of a physician who did. Nonetheless, Wilfred Shute claimed in 1969 that the use of vitamin E in angina "remains largely unknown to the medical profession." This is nonsense. The reality is that it was a treatment proposed with great enthusiasm, but for which no benefit could be shown by others and which, quite properly, was rejected.

In most circumstances, that would have been the end of vitamin E for angina. However, Vogelsang and the Shutes did not go away. They continued to treat their patients with vitamin E and, although they never again presented their results for critical review in the medical literature, they continued to publicize the vitamin through publications of the Shute Foundation for Medical Research as well as in a book by Wilfred Shute entitled *Vitamin E for Healthy and Ail-*

Many physicians, as well as the general public, express dissatisfaction with double-blind, placebo-controlled trials of what they may regard as proven remedies.... several hundred years of experience have demonstrated beyond doubt that careful experimental design is essential if we are to arrive at the truth. Bearing witness...are discarded treatments ranging from bloodletting for pneumonia and inhalation of chlorine gas for respiratory disease in the 18th century to fever therapy for schizophrenia, wholesale tonsillectomy for sore throat, and Laetrile for cancer in the 20th.

ing Hearts. (I found it in my local supermarket right next to *Dr. Atkin's Diet Revolution*.) Largely as a result of this publicity, many thousands of those with angina have continued to take vitamin E either with, or more likely, without the blessings of their physicians. An additional consequence was that after nearly 25 years of inactivity, several more attempts were made to verify the original claims of benefit. Studies that failed to find value in vitamin E were not without fault. Wilfred Shute suggested that inadequate doses were used and that the patients were treated for too short a period of time. To overcome these objections, Dr. Terence Anderson of the University of Toronto consulted the Drs. Shute before beginning his trial of the vitamin. The Shutes suggested that a dose of 3200 IU per day for nine weeks should result in "80% improvement." (Note that the Shutes' recommended dose in 1973, the year of Anderson's study, is ten times larger than in 1947.) Under Dr. Anderson's direction, 36 angina patients were matched for sex and age and randomly assigned to receive either vitamin E or an inert pill (placebo) similar in appearance, taste, and texture. Assessment of the patients was by their personal physicians in a double-blind design; neither the patients nor their doctors were aware who had been assigned to the vitamin E and placebo groups. The need for these elaborate precautions against inadvertent bias may not be obvious. Many physicians, as well as the general public, express dissatisfaction with double-blind, placebo-controlled trials of what they may regard as proven remedies. However, several hundred years of experience have demonstrated beyond doubt that careful experimental design is essential if we are to arrive at the truth. Bearing witness to this fact are very many discarded treatments, ranging from blood-letting for pneumonia and inhalation of chlorine gas for respiratory disease in the 18th century, to fever therapy for schizophrenia, wholesale tonsillectomy for sore throat, and Laetrile for cancer in the 20th.

When the assignment code for Anderson's study was broken, it was found that, in terms of the categories allowed the raters, 13 of the 18 on vitamin E were "unchanged," one was "much improved," and four were "improved." Of those treated with placebo, 12 were "unchanged," three were "improved," two "slightly improved," and one was "slightly worse." How are we to interpret these results? Clearly there is no support for the prediction that 80% would improve with vitamin E. In fact, 72% were unchanged. But what are we to do with the fact that one patient on vitamin E was "much improved" whereas none on placebo were, or the fact that one patient on placebo was "slightly worse," whereas none treated with the vitamin were in that category? The only statistically valid conclusion is that these slight differences in treatment outcome result from chance; if we were to repeat the study several times, the placebo group would look slightly better as often as the vitamin group. However, as Anderson put it, "a small beneficial effect cannot be ruled out."

No amount of reanalysis of Dr. Anderson's data will get us closer to the truth. To use a well-worn phrase, more research is needed. But we must not forget what it is that we

---

The basis for the popularity of many treatments rests on their ability to "cure" a disease...the patient has never had.

---

are unsure of. It is not Shutes' claim that vitamin E is beneficial for most angina patients; we have rejected that claim with great certainty. Instead, we are being tantalized with the possibility of a "small beneficial effect."

The investigation that finally convinced me with respect to vitamin E in angina was conducted by Dr. Ronald Gillilan and his associates at the Public Health Service Hospital in Baltimore and reported in 1977. They were very careful in their selection of patients, requiring that each have typical effort-related angina with evidence of 75% block of at least one major coronary artery.

(The basis for the popularity of many treatments rests on their ability to "cure" a disease that the patient has never had. For example, intoxication with bromide found in drinking water is easily confused with schizophrenia. If a person poisoned with bromide is diagnosed as schizophrenic, placed in a hospital with a source of pure water, and treated with such useless remedies as aloe vera or vitamin E or bee pollen, recovery will occur regardless, and we will have a "new," but almost certainly useless, therapy for "schizophrenia.")

Evaluation of Dr. Gillilan's patients included exercise testing on a treadmill, the number of angina attacks, and the number of nitroglycerin tablets taken. Forty-eight patients were treated with 1600 units of vitamin E per day in a double-blind, crossover study. In a cross-over design, both treatments are given to all patients. Thus, 24 persons were started on vitamin E and 24 on placebo. After six months, the groups were "crossed-over," vitamin E patients to placebo and vice versa. The results of the study were unequivocal. When compared with placebo, vitamin E failed to increase capacity for exercise, failed to decrease the number of angina attacks, and did not alter the amount of nitroglycerin taken.

Whatever your personal decision about the use of vitamin E in angina, I hope you are now convinced of four things:

(1) Claims that vitamin E as a treatment for angina have been ignored by the medical establishment are without basis in fact.

(2) Claims that abundant evidence exists for the efficacy of vitamin E in angina are without basis in fact.

(3) No treatment is ever proven ineffective beyond all doubt, only beyond reasonable doubt.

(4) The ineffectiveness of vitamin E in angina has been proven beyond reasonable doubt.

Publications of the Shute Foundation have also suggested that vitamin E be used to treat high blood pressure, Buerger's disease, gangrene, inflammation of the kidneys, rheumatic fever, the complications of diabetes, atherosclerosis, coronary thrombosis, and varicose veins. If you are fortunate enough to suffer none of these conditions, the Foundation tells us that regular use of vitamin E supplements will prevent stroke and senility. Each of these claims is a separate issue and each could be verified or disproven in the same fashion as I have described for angina. Unfortunately for us, but in favor of the advocates of vitamin E, none of these claims has been or is likely to be evaluated as thoroughly as were its supposed antianginal properties. As a consequence, we can say in all honesty only that these claims are unproven, not that they are disproven. The danger of this uncertainty lies not so much in the use of vitamin E, but in reliance upon it to the extent that other aspects of treatment are neglected. For example, a moderately obese person with adult-onset diabetes who ignores weight reduction and exercise in favor of vitamin E has traded measures of known value for an unproven remedy. In the long run, the consequences of this may be disastrous.

The indignation of the medical establishment regarding the unsubstantiated claims for vitamin E by men such as

---

*Because free radicals are thought to be important in...
aging and in the initiation of cancer, there has been
speculation that dietary antioxidants such as vitamins E
and C, riboflavin, beta-carotene, selenium, zinc, and
copper may have both anti-aging and anticancer properties.*

Wilfred Shute is understandable. Most physicians take seriously their responsibility to protect their patients from unproven and quack remedies. However, reflex rejection of therapeutic claims for vitamin E must be avoided and the same standards of evidence applied to its critics as to its advocates. Alton Ochsner, a distinguished American surgeon, has long recommended the use of vitamin E to prevent the postoperative formation of blood clots in the veins of the legs. The most serious consequence of their formation is that they will break loose and travel to the lungs where blockage of blood flow will result (pulmonary thromboembolism). An equally distinguished American physician, Hyman Roberts, has written in the *Journal of the American Medical Association* that vitamin E can cause pulmonary thromboembolism, the very condition Ochsner believes it prevents. Both are men of honesty and integrity, but neither can present convincing evidence of his position. Nonetheless, each will be quoted selectively by those who are either for or against vitamin E.

As has already been mentioned, vitamin E, by inactivating free radicals, is able to protect unsaturated fats from deterioration. The antioxidant properties of the vitamin form the basis for a few relatively well-established therapeutic uses. The best known of these is treatment of premature infants in an attempt to reduce damage to eyes and lungs that may result from prolonged administration of oxygen. Because free radicals are thought to be important in the processes of aging and in the initiation of cancer, there has been speculation that dietary antioxidants such as vitamins E and C, riboflavin, beta-

> *Unless we wish to play a pharmacological version of Russian roulette, the decision to self-administer any chemical must be based on a realistic appraisal not only of its possible benefits, but also its possible hazards.*

carotene, selenium, zinc, and copper may have both anti-aging and anticancer properties. These possibilities have been given much attention in the popular press and are presently being investigated both in animals and in humans. For example, in 1987 a 5-year study was begun to evaluate the effects of a daily does of 2000 IU of vitamin E in combination with drug treatment on the progression of Parkinson's disease. Readers tempted to begin self-treatment should note that an interim evaluation issued in November 1989 found drug treatment to be very effective, but could detect no benefit from this very large dose of vitamin E.

Unless we wish to play a pharmacological version of Russian roulette, the decision to self-administer any chemical must be based on a realistic appraisal not only of its possible benefits, but also its possible hazards. This relationship is called the benefit–risk ratio and is equally applicable to large doses of an essential nutrient such as vitamin E, a legal drug such as alcohol, or an illicit one such as cocaine. As a general rule, the greater the perceived benefits of a drug, the greater will be the acceptable risks in its use. For example, a drug able to cure a rapidly fatal form of cancer may slightly increase the likelihood of developing leukemia at a later time. In this case, the benefit–risk ratio is favorable. The ratio for a trivial use of the same drug, perhaps for the control of acne, would be completely unacceptable. Establishing a benefit–risk ratio for a substance whose primary effect is to bring pleasure is particularly difficult. How, for example, does one assess the benefit derived from drinking a cup of coffee or eating a cholesterol-rich dessert or smoking a cigaret?

What is the benefit–risk ratio for vitamin E? In those rare instances where deficiency of the vitamin exists, supplementation is of obvious value. However, the ingestion of hundreds or thousands of units of vitamin E per day in the hope that vascular disease will be halted, or one's sex life will improve, or aging will be avoided is quite another matter. For uses such as these, the benefits range from highly unlikely to plausible but unproven. For this reason, we must consider very carefully the possible risks in the use of the vitamin. There is little doubt, as M. K. Horwitt has written, that vitamin E is one of the least toxic of the vitamins.

In a report by the Institute of Food Technologists in 1977, it was concluded that there is no evidence that the tocopherols are toxic even in large doses. The report added however, that "untoward effects are hypothetically possible." At daily doses of 100–300 units, virtually no toxic effects have been reported. In an interesting letter published in the *New England Journal of Medicine* in 1973, H. M. Cohen, a California physician, told of weakness and fatigue in himself and some of his patients taking 800 units per day. However, we must not forget that the placebo effect may be perceived as adverse as well as beneficial. This fact may be illustrated by a study conducted in Mexico in which 147 women were given inert tablets for a year, but were told that they were receiving a contraceptive drug. Thirty different adverse effects were reported, the most common being headache (16%) and a decreased sexual desire (30%). Only one-third of the women failed to report some unwanted effect. (The authors don't mention the reactions of the 72 women who got pregnant during the year.)

Returning to vitamin E, it is noteworthy that in the double-blind studies by Anderson and by Gillilan, there was no difference between the vitamin and placebo in terms of adverse effects even though doses in excess of 1500 units were given. Thus, the risks presented by supplementation of the diet with up to a few hundred units of vitamin E per day appear to be minimal. For this reason, if substantial evidence of even a modest beneficial effect could be presented, an acceptable benefit–risk ratio would exist. From the standpoint of vitamin E as an essential nutrient, it appears impossible for a person who eats a reasonably balanced diet to be deficient. The primary reason for this is the wide distribution of the vitamin in foods. It is found in eggs, liver, fish, dairy products, and fruits and vegetables of all kinds. In addition, vitamin E is relatively resistant to losses during cooking and storage. It is true that an increased consumption of unsaturated fats increases our need, but the vegetable oil margarines, our most common source of such fats, are rich in vitamin E as well. There is no convincing, or even suggestive, evidence that anyone who pays just modest attention to their personal diet needs an additional source of vitamin E.

What advice can be given to those who have decided to take supplemental vitamin E in the hope that one disease or another will be alleviated or avoided? First, it must be reemphasized that the greatest known hazard in using vitamin E is that other means of prevention and treatment will be neglected. A person with high blood pressure who continues to smoke, but takes vitamin E is foolish indeed. But, if we have carefully attended to all other aspects of fitness and health and wish merely to add vitamin E on the chance that it may do some good, we still need to pick a dose. Capsules sold in drug and health food stores typically contain anywhere from 100 to 1000 international units in increments of 100 IU. Should we assume that the more we take, the higher will be the levels of vitamin E in our bodies?

In a study at the National Institutes of Health, Drs. Philip Farrell and John Bieri compared a group of healthy control subjects with a group that had taken supplemental vitamin E in doses of 100–800 units per day for periods ranging from 4 months to 21 years. In 18 control subjects, the average tocopherol level was 650 micrograms per hundred milliliters (range: 550–1100). In those taking vitamin E, the average was about twice as high, 1340, but individual levels were quite variable and ranged from 550 to 2400. If we assume that the controls had an intake equal to the RDA of 12–15 units per day and those taking a supplement an intake of 400 units, we see that a 25-fold increase in intake produced only a doubling of plasma levels. Furthermore, Farrell and Bieri found no relationship between the size of the supplement taken and blood levels of the vitamin; 100–200 units produced exactly the same level as did 600–800 units. These observations indicate that the human body regulates the level of vitamin E rather closely. Beyond a certain level, additional amounts will simply be eliminated in the feces. Such a mechanism would also account for the low toxicity of the vitamin; most of a very large dose would be eliminated before it could do harm. It appears that the blood level of vitamin E provided by a normal diet can be doubled by supplementation and a hundred units are as likely to do this as are much larger doses.

Summing Up

Vitamin E remains an intriguing substance. The evidence before us clearly indicates that deficiency simply does not occur in normal humans. How then are we to explain the enormous popularity of vitamin E supplements for the American people? This chapter has provided some answers to that question: a variety of interesting effects on animals deprived of the vitamin, widespread promotion of the unsubstantiated claim that vitamin E is useful in treating heart disease, accepted value in preventing oxygen toxicity in premature infants, unrelenting advertisements by the vitamin industry, and, perhaps most important of all, an absence of toxicity even at doses enormously in excess of the RDA. Nothing there to convince me to go outside my diet for vitamin E. Yet, in my mind, lingers the thought that there are importants things yet to be discovered about vitamin E; perhaps life is but a race against the rancidity of my unsaturated fats and E will give me an edge. For now I'll stick with whole grains and lettuce, and leave the supplements on the shelf. For the future, as Max Horwitt expressed it in 1989, "Let the debate begin!"

CHAPTER 17

Calcium

Diet Factor of the Year

---

*Magazines...are filled with advertisements for calcium
supplements. Dairy Councils...tell us that the calcium
in milk will surely prevent high blood pressure, cancer,
and osteoporosis.... I will tell you now that 99%
of the claims made for calcium supplements range
from misleading to downright wrong.*

In these days of award shows, I occasionally imagine a
glamorous ceremony called the "Pop Nutrition Hall of Fame
Dietary Factory of the Year Award." Magazines, especially
"women's" magazines, are filled with advertisements for
calcium supplements. Dairy Councils across the country tell
us that the calcium in milk will surely prevent high blood
pressure, cancer, and osteoporosis. Products like Rolaids and
Tums, promoted for years as antacids, have suddenly become
"calcium supplements" as well.

D. Mark Hegsted is Professor Emeritus of Nutrition at
the Harvard University School of Public Health. Based on
his studies of human subjects, he has said that "the minimum calcium requirement of adult males is so low that deficiency is unlikely on most natural diets." Alexander R. P.
Walker is the head of the Human Biochemistry Research Unit
of the South African Institute for Medical Research. He is,
like Hegsted, a nutritionist of international stature. We earlier met Dr. Walker in Chapter 3 as a leading advocate of
increased fiber in the diet. Of calcium, Dr. Walker says "there

is no firm evidence that calcium deficiency exists in humans...it is questionable whether calcium merits a place in the tables of recommended allowances of nutrients."

Can the "calcium" of Walker and Hegsted possibly be the same "calcium" in the advertisements? Yes. To understand how we've gotten from "no firm evidence that calcium deficiency exists in humans" to the current flood of advice to take calcium supplements will require a little patience. We began the story in Chapter 15 with vitamin D and rickets; I will continue it in this chapter with a general discussion of the RDA for calcium and how that requirement is best met; and we will finish in Chapter 24 by considering exercise, calcium, and the rest of the diet as factors in osteoporosis. But this is not a mystery novel that I write; I will tell you now that 99% of the claims made for calcium supplements range from misleading to downright wrong.

We will begin by recalling a few things about vitamin D and the disease called rickets. It required no great imagination to put together the fact that bones are made of calcium and the fact that rachitic children have too little bone, to reach the conclusion that the disease would be cured by adding calcium to the diet. And it required no great effort to prove that conclusion wrong. Mellanby's puppies and the children of the smoky, sunless cities of Europe were devastated by rickets, not because there was too little calcium in their food, but because, in the absence of sunlight-generated vitamin D, too little calcium was absorbed from the food passing through their guts. If vitamin D is present, diets miserable by any standard appear to provide enough calcium to avoid rickets. Conversely, no amount of dietary calcium is adequate if the body, because of vitamin D deficiency, is unable to absorb it.

In 1920, W. C. Sherman of Columbia University proposed that the human need for calcium is about 450 mg per day. He noted, however, that there seems to be considerable variation between individuals and that some might need much more than the average. In general, subsequent investigators confirmed Sherman's findings and, in 1943, the first edition of the Recommended Dietary Allowances set the RDA at 800 mg. In the tenth (1989) edition, that value remains unchanged for men and women 25 years of age and older. However, it is now suggested that from age 11 to 24, the period of major skeletal growth, calcium intake be increased to 1200 mg. The relative stability of these values is deceptive; it masks a controversy over our need for dietary calcium that has raged for the past half century and shows no signs of ending.

First we will consider the evidence of people like Hegsted and Walker, representatives of what might be called the minimalist school. In the late 1940s, Professor Hegsted was concerned about the low level of calcium in the diets of many of the people of South America. In most countries, dairy products, and to a lesser extent green leafy vegetables, are the primary sources of calcium. For many in tropical and subtropical countries, especially the poor, these sources may be virtually absent from the diet.

Hegsted's subjects were ten male inmates of the Central Penitentiary of Lima. They had been in prison for from two to 20 years and seemed well-adjusted to the food provided them. Because of an almost total lack of foods rich in calcium, their intake of the mineral was estimated to be 100–200 mg per day, only a fraction of the recommended allowance in this country. In his experiments, Dr. Hegsted varied the amount of calcium in the diet and carefully measured the amount the men excreted. These data indicated an average requirement of 126 mg per day. Hegsted's conclusion was pragmatic: Don't worry about the calcium intake of adult males; available calcium should be reserved for children and women of childbearing age. (In 1986, Hegsted proposed 300–400 mg per day as a reasonable calcium RDA for Americans.)

Like Professor Hegsted, A. R. P. Walker began his studies of calcium in the 1940s and, like Hegsted, he has continued with them for nearly four decades; these men are not dabblers in nutrition. The people of greatest interest to Walker

---

...there has been constant tension between what I have
called the calcium minimalists and the calcium maximalists.

---

are the natives of south and central Africa, especially the South
African Bantu. In comparison with Europeans living in South
Africa, the Bantu intake of calcium is quite low. Walker estimates the range to be 175–475 mg per day. Despite this difference, rickets and osteomalacia are as uncommon in blacks
as in whites. With respect to osteoporosis, elderly Bantu are
much less likely to suffer spontaneous fractures than are the
Caucasians of South Africa. Walker's conclusions are simple
enough: Low intake of calcium is not harmful to humans,
and an increase in calcium intake is unlikely to be beneficial.

Now let's turn from the minimalists to the calcium maximalists; today their influence is far more in evidence in the
popular media. During the years that people like Hegsted
and Walker were gathering their data, others were busy as
well. F. R. Steggerda and H. H. Mitchell at the University of
Illinois studied American males who were accustomed to a
high calcium diet. In these individuals, the apparent need
for calcium was much greater than that of Hegsted's prisoners or the Bantu. The average value was 700 mg. Furthermore, there was considerable variation between individuals.

Using standard statistical methods, Steggerda and
Mitchell estimated that the calcium needs of 99% of the population would be included in the range from 340 to 1060 mg
per day. Because individuals have no way of knowing where
they are in that range and because there is no apparent harm
in a moderate excess of calcium, many authorities took the
position that everyone should be provided by their diet with
at least 1000 mg per day. That position was formalized in
the 1946 edition of the Recommended Dietary Allowances;
the RDA for calcium was increased to 1000 mg.

In the years since the second edition of the RDAs was
issued there has been constant tension between what I have

called the calcium minimalists and the calcium maximalists. Until only a few years ago, the tide was running very slightly in favor of the minimalists. For example, the World Health Organization could find no relationship between bone health and calcium intake in the countries of the world and suggested an RDA of 400–500 mg. Those responsible for the American RDAs reduced the RDA from 1000 mg back to its 1943 level of 800 mg. At the same time recognition was taken of the peculiar needs of certain groups. Thus, in 1989, we have the recommendation for 1200 mg of calcium per day for pregnant women and nursing mothers, as well as for those age 11 to 24.

Despite what I've told you about the work of scientists like Hegsted and Walker, and about the epidemiological investigations that have repeatedly failed to find a relationship between bone health or any other kind of health and the intake of calcium, the RDA is not about to be lowered to 400 or 500 mg. Despite Hegsted's "strenuous objections to dietary standards which force the conclusion that thirty percent of the US population is not getting enough calcium," the Tenth Edition of the Recommended Dietary Allowances actually increases the amount suggested for men and women between the ages of 19–24.

Several reasons can be offered for the swing toward the calcium maximalists. Primary among them is the disease called osteoporosis, a disease of unknown origin that affects at least one-quarter of the elderly. It is a disease that is more prevalent in women. It is a disease for which no cure is known.

In 1984 a panel of experts on osteoporosis was called together by the National Institutes of Health. After several days of deliberation, their recommendation was that women increase their intake of calcium. Until a cure is found for osteoporosis, that recommendation will insure the continued health of companies with calcium supplements to sell. Whether that recommendation will have any influence on the frequency or severity of osteoporosis remains to be seen.

We shall return to osteoporosis in Chapter 24.

I don't know how much calcium Professor Hegsted or Dr. Walker consume. I do know that they have convinced me that our bodies can adapt to lowered intake of calcium simply by increasing the fraction that is absorbed and by decreasing the amount that normally is lost in the urine. The millions of people around the world who have perfectly sound bones on a lifetime intake of a few hundred milligrams of calcium a day provide impressive evidence of that. But in this instance I choose not to practice what they preach. In an affluent society in which natural sources of calcium are readily available, I believe it reasonable for all men and nonpregnant women to ingest about 1000 mg of calcium each day. At this level, the benefits of high calcium intake, if there be any, will be realized and, with minor exceptions to which we shall come, there is no known toxicity. Just how easy it is to obtain 1000 mg of calcium without excess calories and saturated fat, and without resorting to supplements, may surprise you.

The usual way in which calcium intake in a country is estimated is not by the study of individual diets, but by checking the health of the dairy industry. For example, in the United States, milk, cheese, and other dairy products provide more than 60% of our calcium. The African and Asian countries, where typical daily calcium intakes are in the range of a few hundred milligrams, are areas in which for religious or economic reasons milk is seldom consumed.

A resident of Burbank, California who drinks a quart of milk and a quart of city water each day has a calcium intake of nearly 1300 mg. The milk provides 1200 mg and Burbank's rather hard water another 80 mg or so. The total is 160 percent of the adult RDA for calcium without even considering other dietary sources.

With this example before us, we may well wonder what all the talk of our need for calcium supplements is about.

Milk and milk products are a very convenient source of calcium. If you are not concerned about fat or calories in your diet, any milk product will do as far as calcium is concerned. For the vast majority of adults in this country interested in limiting their fat and caloric intake, a little label-reading is in order. The 1200 milligrams of calcium in a quart of skim milk carries with it only 360 calories. A quart of whole milk has no more calcium, but more than twice the number of calories. Most cheeses have an even worse calcium/calorie ratio. Furthermore, the difference in caloric content is almost entirely in the form of saturated fat.

Many Americans, particulary those of African descent, are intolerant to lactose, the sugar found in milk. For them, yogurt may provide a palatable alternative source of calcium. Again, read the label; low-fat and even no-fat yogurts are available if you're concerned about calories. One other thing about yogurt: Unlike the milk you drink, yogurt usually is not supplemented with vitamin D. For persons who consume no vitamin D-supplemented milk and are habitually denied sunlight because of confinement indoors or by the regular wearing of restrictive clothing, a modest vitamin D supplement is in order.

I mentioned drinking water as a source of calcium. The "hardness" of water is largely determined by its content of calcium carbonate. In parts of this country and the world where the water is especially hard, 100 mg or more of calcium may be obtained just by drinking a quart of the water. For many years there has been argument among epidemiologists about a possible relationship between water hardness and the incidence of heart disease. Perhaps hard water has some protective effect and perhaps its calcium content is important in this regard.

With all this talk about dairy products as rich sources of calcium, we might expect to find calcium deficiency rampant among strict vegetarians (vegans), persons who consume no animal products. In fact, adequate amounts of calcium can be provided by vegetables alone. The reason is that calcium is found in nearly all plants; cows, you may recall, don't drink milk. A study done a few years ago in Sweden found that a small group of vegans had an intake of calcium only 10% lower than that of persons eating a traditional Swedish diet.

Although all vegetables contain some calcium, only a few can be thought of as calcium-rich. These are dandelion, turnip, and mustard greens, kale, collard, rhubarb, and several others. If you have a taste for such things, they represent a low calorie source of calcium that contains no saturated fat. One should not conclude from what I've just said that just any vegan diet will provide adequate calcium. When all dairy products are avoided there is a very real risk, especially for infants, children, adolescents, and pregnant women, that too little calcium will be left in the diet.

Other than dairy products and the vegetables that I've mentioned, not many foods can be regarded as reliable sources of calcium. Oysters and bony fish such as sardines and salmon are high in calcium but, because of expense or availability, few eat them with regularity. A cup of almonds provides about as much calcium as a cup of skim milk; unfortunately the almonds cost four or five dollars a pound and contain ten times the number of calories. Many have the idea that red meat is a good source of calcium. This is not only not true, but there is some evidence that high meat diets impair calcium balance.

I already have said that vitamin D is of overwhelming importance to the absorption of calcium. That fact is often lost in a maze of confusing statements about dietary phytate, phosphorus/calcium ratios, and high protein as factors in calcium absorption. That phytate, phosphorus, and protein can influence calcium balance is without question; that they are significant factors remains to be established. At the present time the evidence with respect to protein and phosphorus is too contradictory to permit any firm recommendation; phytate is a simpler matter.

In the discussion of dietary fiber in Chapter 3, phytic acid was mentioned as a substance that binds minerals such as iron, zinc, magnesium, and calcium. Whole wheat breads, for example, may be less nutritious than their mineral content would suggest because phytic acid renders a portion of the minerals unabsorbable. The overall importance of phytic acid in calcium balance is uncertain. Those following a strict vegan diet certainly should be concerned about the phytate content of their vegetable sources of calcium. For those whose primary source of calcium is dairy products, phytic acid surely is insignificant.

At several points in this book I have expressed my opposition to the uncritical use of dietary supplements. Any healthy adult able to read this sentence is able to arrive at a completely adequate nonsupplemented diet. For most of us, calcium is no exception; a pint of skim milk alone will provide 600 mg. But what of those who don't like milk, or who have allergies to its proteins, or who flat out refuse to think about what they eat; perhaps a pregnant teenager with an RDA of 1200 mg? A calcium supplement may then be in order.

The forms of calcium most often found in health food stores and vitamin catalogs are the carbonate, lactate, and gluconate salts, bone meal, and something called dolomite. A few advertise ground-up oyster shells. The Food and Drug Administration has warned us of the possibility of lead contamination in bones, so I would be inclined to avoid any "natural" sources of calcium, especially if the supplement use is to be long term.

Studies that compared the carbonate, lactate, and gluconate salts have found only minor differences between them. Preliminary evidence suggests that a fourth salt, calcium citrate, may have some advantages in terms of ease of absorption and lesser probability of stone formation. If you should choose to use an antacid as a source of calcium, be sure to read the label carefully. Rather than calcium carbonate, some contain aluminum salts that may actually deplete calcium.

With vitamin catalogs offering us 1000 mg per tablet calcium supplements, it's not hard to imagine large num-

---

*If we are lucky, today's calcium craze will not be
followed by an epidemic of kidney stones*

---

bers of people, especially women freightened by the spectre
of osteoporosis, consuming enormous amounts of calcium.
What kinds of toxic effects might we expect to see? The most
likely adverse consequence is an increase in the incidence of
kidney stones. The magnitude of the risk for completely normal individuals is unknown, but those with a history of stone
formation should certainly be wary of calcium supplements
and, for that matter, of the water they drink. The advice to
drink lots of water in order to avoid stones is valid only if it's
not hard water. A few years ago in Israel, where stone formation is a major problem, some patients were found to be
getting as much as a 1000 mg a day from the water they drank.

By and large, my body is much better able to handle large
doses of calcium than it is large doses of vitamin D; most
excess calcium simply goes down the toilet. If we are lucky,
today's calcium craze will not be followed by an epidemic of
kidney stones or by some other unforeseen calamity.

Summing Up

Evidence from around the world is abundant that, given
an adequate supply of vitamin D, true calcium deficiency is
rare indeed; several hundred milligrams per day of the mineral is adequate for most of us. Thus, it is hope, rather than
convincing evidence of need, that sustains our present RDA
of 1200 mg of calcium for those age 11–24 years, hope that an
excess of calcium during the time of maximal skeletal growth
will protect against the disease of osteoporosis many years
later. Some authorities, perhaps angered by the flagrant
deceptions of those who sell calcium supplements, have chosen to dash that hope. I prefer to encourage, at least until a
cure for osteoporosis is found, a generous intake of calcium
in the form of skim milk and other no-fat dairy products.

CHAPTER 18

Iron

Tonic or Toxin?

...anemia caused by lack of iron is the most common deficiency disease in the world.

Anemia is a condition in which the blood is unable to carry sufficient oxygen from the air we breath to the tissues of our bodies, as mentioned earlier (Chapter 12). Among the consequences of severe anemia are pallor of the skin, shortness of breath, extreme weakness, heart palpitations, and death. In contrast, the symptoms of mild anemia may be so vague as to be ascribed to boredom, laziness, or neurosis. The three nutritional factors most commonly associated with anemia are vitamin B_{12} (Chapter 12), folic acid (Chapter 13), and iron. Of the three, iron is numerically the most important. It often is said that anemia caused by lack of iron is the most common deficiency disease in the world.

Iron has a long history as a medicinal agent; 1500 years before the birth of Christ, the physician–priests of Egypt prescribed iron. But it was not until quite recently that the central role of iron in the prevention of anemia was fully appreciated. Oddly enough, because of the widespread and longstanding practice of drawing blood from their patients, physicians over the centuries have been more likely to cause anemia than to prevent it; John Burnum calls them "medical vampires."

Throughout this book we have seen numerous examples of plausible ideas that have turned out to be just plain wrong.

225

Both the idea that we become ill because of bad things in our blood and the idea that we might become well by removing the bad blood certainly sound like common sense; George Washington is reported to have had a quart of blood drained off on the day he died. Such bloodletting is not much done these days. On the other hand, the curious notion that a mild degree of anemia might be good for some of us some of the time is still prevalent; curiouser still, it may be correct. But let us examine simpler things first.

Iron is essential for all forms of life. In the human body, most of the iron is found in the blood in the form of hemoglobin, the molecule that gives red blood cells the ability to carry oxygen. The barrier between the blood pumped through the lungs and the air we breath is but a single layer of cells. Oxygen diffuses across that layer, combines with hemoglobin, and the two are carried to all cells of the body where the oxygen is released.

Another gas with which hemoglobin combines is carbon monoxide. In fact, hemoglobin so much prefers carbon monoxide over oxygen that even a small amount of carbon monoxide in the air can cause oxygen deficiency in our tissues. Together with all of the other hazards of their addiction, smokers of cigarets are chronically exposed to low levels of carbon monoxide. The resulting deficiency of oxygen may be modest in degree and subtle in its consequences. If, on the other hand, I share a closed garage with a running automobile engine, carbon monoxide will quickly be lethal. Iron deficiency anemia, like carbon monoxide poisoning, comes in all degrees of severity, from the lethal to the subtle.

Healthy adult males and postmenopausal females are the segments of the population least likely to suffer iron deficiency anemia. The reasons for this fact tell us much about the iron economy of the body. Mature men and postmenopausal women do not become pregnant, do not grow, and do not regularly bleed. When their red blood cells become old at the age of 120 days or so, they are destroyed, but the iron is

> *...when we grow, the volume of blood grows as well, and with it the need for iron. It is for this reason that iron deficiency anemia is common in children, unborn babies, and the women who carry them.*

saved, sent to the bone marrow, and made into the hemoglobin for new cells. Whatever tiny losses of iron that may occur are easily made up by absorption of iron from the diet. Anemia in postmenopausal women and adult men is most often a sign of hidden bleeding. Identification of the cause of blood loss is far more important than the anemia that results.

When thinking about growth, such ideas as protein for muscle and calcium for bone come to mind. But when we grow, the volume of blood grows as well, and with it the need for iron. It is for this reason that iron deficiency anemia is common in children, unborn babies, and the women who carry them. Women of childbearing age have an additional burden in the regular loss of blood during menstruation.

The differences in demand for iron by adult men, growing babies and children, and menstruating women are reflected in the Recommended Dietary Allowances for iron. The needs of children up to the age of ten years, of males 19 years of age and beyond, and of women who have reached the age of menopause are at the base: 10 mg per day. The increased demand for iron during growth is reflected in an RDA of 12 mg for males between the ages of 11 and 18 years. Because females not only must grow, but must contend as well with the regular loss of blood during menstruation, the RDA is 15 mg at age 11 and stays at that level until menopause is reached. Finally, women are advised to take in 30 mg of iron per day during pregnancy.

The apparent simplicity of the RDAs for iron is misleading. There is today continuing uncertainty regarding the wisdom of the specific values chosen and what should be done about meeting them. Indeed, the 1989 RDAs reflect a decrease from the 1980 edition in the amount of iron recommended in 12 of the 18 categories listed and in no category is the RDA increased. These revisions were based primarily on two factors. The first was increased concern with the toxicity of iron. The second was the recognition that the amount of iron absorbed from the diet increases as iron stores decrease.

Two facts about iron are at the root of the problem facing nutrition scientists. The first has to do with the influence of complex dietary and physiological factors on iron absorption from the gut. For example, human breast milk provides only about a quarter of a milligram of iron per day, but fully half of it is absorbed by infants. In contrast, some foods are rich in iron, but little of it is absorbed. Thus the RDA of 10 mg for an adult male is based upon a mixed diet containing iron of varying absorbability; actually, somewhat less than a milligram of iron is expected to be taken up by the gut.

A simple answer to the problem of providing adequate dietary iron is to add excess iron to a variety of foods; in fact, this approach often is advocated and often is followed. For example, in the United States, iron-enriched flour accounts for about 20% of dietary iron. Which brings us to the second fact about iron: in overdose, it is a very toxic substance.

Iron poisoning takes place in two ways. The more readily understood is the simple overdose. It is true that the normal human gut is able to reject large amounts of iron in the diet and thus to protect most of us most of the time from being poisoned. It is true as well that the barriers to absorption can be overwhelmed. A classic example is provided by the Bantu of South Africa.

Shortly after steel drums were introduced into South Africa by European settlers, the native Bantu began to use old drums as containers for brewing beer. Over a period of years, Bantu beer drinkers were found to have a very high incidence of cirrhosis of the liver, the condition that kills so many alcoholics. But the primary problem was not alcohol;

---

*Given the vigorous promotion and availability
of iron supplements in this country, it is remarkable
that so little iron poisoning is reported.*

---

it was an excess of dietary iron, iron leached from the drums during the brewing process.

Given the vigorous promotion and availability of iron supplements in this country, it is remarkable that so little iron poisoning is reported. It appears that most of us can exceed the RDA for iron by severalfold without causing obvious harm; this is not a recommendation for doing so. A recent advertisement from one of the national sellers of vitamins proudly proclaims that just four of their iron pills will provide 639% of the RDA for iron. And a Rodale book called *The Healing Power of Nutrition* promises to tell us "How to pump up the iron content of your spaghetti sauce to *more than 8 times the RDA*" (their emphasis). No reasons for wishing to ingest large excesses of a potentially poisonous substance are given by either company.

The most frequent victims of poisoning by iron are young children who eat a bottle or so of an iron-fortified children's vitamin. (Toddlers find almost irresistible a bottle decorated with Bugs Bunny or the Flintstones.) The lessons here are that vitamins and minerals in excess may be toxic chemicals and that, almost all the time, good food is a better, safer, and more pleasurable source of nutrients than pills.

Iron overdose would be a simple matter if it were dependent only upon iron intake. But, as is true for so many other aspects of our lives, genetic factors play a role as well. One in 20 of us carry a gene that influences the absorption of iron; Victor Herbert has called it the "iron-loading gene." For the one in 200 who receive the iron-loading gene from both parents, iron poisoning is an ever present danger. The disease that results is called hemochromatosis and its treatment, odd as it may seem, is periodic bleeding of the patient.

---

*Among the consequences of untreated
hemochromatosis are cirrhosis of the liver, diabetes,
heart failure, increased susceptibility to infection,
and impotence in middle age.
The promoters of iron as a tonic for all that ails us
may wish to give hemochromatosis a thought.*

Among the consequences of untreated hemochromatosis are cirrhosis of the liver, diabetes, heart failure, increased susceptibility to infection, and impotence in middle age. The promoters of iron as a tonic for all that ails us may wish to give hemochromatosis a thought.

Against this background of irregular absorption and a high degree of toxicity, the RDAs for iron are seen to be little more than crude guidelines based upon a number of rather tenuous assumptions. Early in this chapter was a statement, the truth of which I do not doubt, that anemia caused by lack of iron is the most common deficiency disease in the world. But widespread anemia is not diminished by an overdose of iron for you and me any more than my getting fat will reduce caloric starvation in Ethiopia.

William H. Crosby, MD, a respected authority on the anemias, is of the opinion that "iron deficiency continues to be overplayed as a public health threat and hemochromatosis, a real killer, is played down." Professor Eugene Weinberg of Indiana University reminds us that, in Sweden, a doubling of iron fortification of flour between 1962 and 1970 was followed in less than a decade by a 350% increase in liver cancer in Swedish women. Cause and effect?—just perhaps. Even the universal advice that pregnant women take an iron supplement has been questioned. After a thorough airing of the controversy in the *British Medical Journal*, Peter Rubin, a professor of therapeutics, concluded that "wholesale supplementation is probably inappropriate." [Having quoted Dr.

Rubin, I must add that most other experts do not agree with him. The consensus remains that pregnant women should get 30 mg of iron per day and, for most, that will require the addition of a modest supplement to their usual diet.]

In all of this, the therapeutic principle is simple: Do not treat a disease [anemia] with a potentially toxic agent [iron] unless you are reasonably certain that the disease is present. Iron-deficiency anemia can be diagnosed only by a competent physician using appropriate laboratory tests. You have been warned: Excess iron may be hazardous to your health. Now we may turn to the pleasant task of assuring an adequate intake of iron from our foods.

An obvious way to insure enough iron for our own blood is to drink the blood of other animals. As attractive as that prospect may be to Count Dracula, most of us prefer to be more subtle. We take our blood in the form of animal flesh. A mere 3.5 ounces of beef contain about 6 mg of iron. Furthermore, blood-iron or, as it more formally is called, heme iron, is very well absorbed by the human gut; depending upon our need for the mineral, as much as 25% of heme iron is taken up. Though heme iron may be only 10% of the total iron in a nonvegetarian diet, it may account for a third of the iron absorbed. As a further bonus, the presence of small amounts of meat or fish in the diet enhances the absorption of nonheme iron, that found in grains, fruits, and vegetables. It is for these reasons that otherwise healthy people who consume meat several times a week are assured an adequate supply of iron.

The fact that meat-eaters are well supplied with iron does not mean that even strict vegetarians need be anemic. Indeed, surveys of vegetarians who consume a variety of fruits,

---

...widespread anemia is not diminished by an overdose of iron for you and me any more than my getting fat will reduce caloric starvation in Ethiopia.

---

*...the effects of iron-deficiency anemia may be quite
subtle....two have gained most attention—
impaired intellectual development in children
and diminished athletic performance.*

---

vegetables, and grains find no evidence of iron deficiency.
On the other hand, vegetarians in general, and women and
children in particular, must be more careful in their selection of foods and in their combinations of foods.

The single most important determinant of iron absorption
from vegetarian meals is the vitamin C content; vitamin C
(ascorbic acid) acts as a promoter of the absorption of nonheme iron. For example, a 1-oz serving of an iron-fortified
cereal such as Kellogg's Fruit and Fiber contains about 10
mg of iron. When eaten as part of a breakfast that includes 8
oz of orange juice, that single ounce of cereal provides enough
iron for the whole day. Without orange juice, iron absorption
from the cereal would still be about right for an adult male,
but inadequate for a woman of childbearing age.

A second factor to be considered by those who eat little
or no meat or fish is the presence in the diet of inhibitors of
iron absorption. Though much has been made of the inhibi-
tory effects of fiber and phytate, tannins found in tea and
coffee are at least as important. Even in the presence of
ascorbic acid, the absorption of nonheme iron may be decreased by 50% or more. For this reason, women and children who habitually follow a vegetarian diet should not drink
tea or coffee with or soon after their meals.

Earlier I mentioned that the effects of iron-deficiency
anemia may be quite subtle. In fact, there is reason to believe
that a number of the consequences of a lack of iron are quite
independent of hemoglobin production and the oxygen carrying capacity of the blood. This thought makes sense in
terms of what we know about the diverse functions of iron.

It is not only a part of hemoglobin; it serves as well in myoglobin, the form of hemoglobin found in muscle fibers, and in a variety of enzymes, the biological catalysts essential for cellular function. The potential consequences of non-anemic iron deficiency are not fully understood, but two have gained most attention—impaired intellectual development in children and diminished athletic performance.

Two decades have passed since the first reports appeared suggesting that iron deficiency even in the absence of anemia might impair the development of the human brain. In the intervening years there have been a number of investigations around the world; children of the poor and underprivileged are the usual subjects. The results to this day are inconclusive; there are too many confounding variables in the life of a child raised in poverty and the instruments for measurement of intellectual, emotional, and motor development are too crude.

In various places in this book I argue against nutritional programs based on inadequate scientific evidence. In this chapter I have pointed out the dangers of excess dietary iron.

Nonetheless, attempts to avoid iron deficiency in pregnant women and in children are so likely to be of benefit and so unlikely to do harm that action is warranted. A model for intervention is provided in this country by the Special Supplemental Food Program for Women, Infants, and Children (WIC). In place since 1973, the program provides iron-fortified formula from birth to 12 months and iron-fortified cereal and vitamin C-enriched fruit juices beginning at age 6 months to children of the poor. The enriched cereals and juices are continued from ages 1–5 years with the addition of milk and eggs.

The efficacy of WIC is indicated by the virtual elimination of iron-deficiency anemia and the less severe stages of iron deficiency in children served by the program. The continuing tragedy is that less than half of the eligible women and children are reached by WIC. Those fortunate enough to live above the poverty level cannot directly participate in WIC, but its dietary guidelines represent good advice for all infants, young children, and pregnant women.

Concern about the iron status of infants and children is well placed and deserves our attention as individuals and as a society. Of much less importance to all but athletes and their supporters is the role of iron in athletic performance. In Chapter 19 we will see that a very significant part of the aerobic conditioning effect is an increased ability to deliver oxygen to exercising muscles. Given the fact that hemoglobin is the carrier of that oxygen, would not an increase in hemoglobin concentration be accompanied by an increase in aerobic capacity? The answer is yes and "blood doping" is the name given to the procedures by which hemoglobin is increased. Simply put, blood is drawn from an athlete well in advance of competition and the hemoglobin-containing red blood cells are concentrated and stored. Then, just prior to competition, the cells are reinfused. Blood doping was as much a part of the 1988 Olympics as were Ben Johnson's anabolic steroids.

Given the well-established negative effects of anemia on performance and the nearly as well-established positive effects of blood doping, many athletes have been attracted to the notion that their less-than-world-class performance might be the result of "iron-poor blood." Because the talents of most such athletes do not warrant the transfusion team and attending hematologist and cardiologist required for safe blood doping, the temptation is to load up with iron supplements. Certainly a plausible idea and one that is understandably attractive both to athletes and to those with iron supplements for sale.

Because the relationship of iron to athletic achievement appears to so many to be so obvious, it may be worth our time to look at two reports published in February of 1988. One study was conducted in Massachusetts, the other in Texas; both focused upon young female athletes, the segment of the population we would predict to be most susceptible to depletion of iron; both attempted to relate non-anemic iron deficiency to performance.

Examination of bone marrow is the most reliable way to estimate iron stores. Unfortunately, getting a sample of the marrow requires an unpleasant procedure to which few other than exercise physiologists and their graduate students would casually submit. As an alternative, measurements are made of a protein in blood called ferritin. It is as ferritin that much of the body's iron is stored within cells. Early in the 1970s it was observed that the amount of ferritin found in the serum provides a reliable index of the amount of iron in the cells. Today, measurements of serum ferritin (SF) are routinely used as an indication of iron stores.

Thomas W. Rowland, MD, and his colleagues measured serum ferritin in members of the girl's cross-country teams in schools around Springfield, Massachusetts. Of 30 girls tested, none were anemic, but 14 were found to have low iron stores, having SF values of less than 20 micrograms per liter.

At the start of Dr. Rowland's study, the girls' endurance was measured by running to exhaustion on a treadmill. These tests were repeated after 4 and 8 weeks. However, after the second treadmill run, seven of the girls took daily iron pills while the others were given placebo. SF measurements at the end of 4 weeks of supplementation or placebo indicated a further drop in SF in the placebo-treated girls, whereas those receiving iron had a significant increase in SF. In terms of endurance, all seven treated with placebo had declines, while six of seven supplemented with iron improved their endurance. The authors concluded that "the results support the hypothesis that nonanemic iron deficiency is deleterious to health." An editorial comment by William B. Strong, MD, head of the sports medicine section of the *American Journal of Diseases of Children*, went further: "The data...continue to support the thesis that nonanemic iron-deficient female adolescent runners can improve their athletic performance by correcting iron deficiency."

Plausible as are the statements by Drs. Strong and Rowland, we must remind ourselves that no measurements of "health" were made and that, beyond treadmill running, no measurements of "athletic performance" were made. Indeed, what is reported is perhaps less interesting than what is left out. Despite the fact that the study was conducted in the middle of a competitive season, we are told nothing of the correlation between iron stores and the results of races in which the girls participated. Such data are so readily obtained and so beautifully quantified that their absence is curious. I am reminded of a report a few years ago by Dr. Johanna Lampe and her colleagues of the University of Minnesota. Nine female marathon runners were evaluated during 11 weeks of training for a marathon. Eight of the nine had average SF values less than 50 micrograms per liter. The observation that the woman with the lowest average value in the group, 7 micrograms per liter, had the third fastest time of the group and set a personal record by 4 minutes suggests that there may be a few things yet to be learned about the relationship between iron stores and performance.

At about the same time that Dr. Rowland and his associates were conducting their study, William L. Risser, MD, and his colleagues at the University of Texas Medical School in Houston compared 100 female intercollegiate athletes with 66 non-athletes. The athletes competed in a variety of sports, all at the NCAA Division I level. Based on a criterion of SF less than 12 micrograms per liter, depletion of iron stores was found in 31% of the athletes and in 46% of the non-athletes. The estimated average dietary iron intake was less than the RDA of 18 mg per day, but there was no correlation between intake and SF status.

Initial screening took place early in the competitive season. Blood measurements were made and evaluations of mood and performance were asked of both the women athletes and their coaches. All of the athletes with serum ferritin values less than 16 micrograms per liter were given a

> *The usual responses to those who urge caution*
> *in the use of vitamin and mineral supplements*
> *include..."supplements provide nutritional*
> *insurance," "they can't do any harm," and "if a*
> *little is good, more is better." With respect to iron*
> *supplements, we know that harm can be done...*

daily supplement containing 65 mg of iron; all others received identical capsules that contained no iron; neither the women nor their coaches were told of iron status or of the treatment assignments. At the end of the season, blood iron, mood, and performance were assessed as before.

In contrast with some earlier studies that found no effect of iron supplements on the iron status of athletes, Dr. Risser's group observed a general increase in iron stores following 3 months' use of the supplement. However, neither the women who received supplemental iron nor their coaches reported "a greater improvement in performance or mood than athletes receiving a placebo." This finding is doubly damning of the notion that iron is a panacea for the woes of athletes. Not only did iron not improve performance, it did not improve performance in women judged by modern methods and criteria to be deficient in iron stores. The authors put it this way: "From these results one might conclude that mild iron deficiency has little effect on elite athletes' performance and therefore need not be identified and treated." They go on to express the caveat, with which I think no reasonable person would disagree, that "more objective means of evaluating competitive performance must be developed before this conclusion is accepted."

The usual responses to those who urge caution in the use of vitamin and mineral supplements include statements such as "supplements provide nutritional insurance," "they can't do any harm," and "if a little is good, more is better."

With respect to iron supplements, we know that harm can be done in persons genetically disposed to storage of excess iron and the development of hemochromatosis. But are there effects more subtle, effects that might occur in all persons who take out "nutritional insurance" in the form of, let us say, 100 mg of supplemental iron each day? I have no certain answer to that question, but there are a number of possibilities worth thinking about. Some of these are simple: Might excess iron contribute to an imbalance of other minerals? It probably can; possible zinc deficiency is most often mentioned. Others are more complex and hypothetical, such as possible effects on infection and immunity and cancer.

Early in this chapter, we stated that "iron is essential for all forms of life." Might I now reasonably conclude that iron is good everywhere and always? Perhaps so, but I must remind myself that some forms of life (mine, for example) are more important than others. Upon invasion by bacteria, our bodies respond in many ways. One is to reduce the amount of iron in the blood; less is absorbed from the gut and more is stored in forms inaccessible to invading bacteria. We become feverish, a response now thought, within limits, to be beneficial. Oddly enough, one of the effects of fever is to decrease the ability of bacteria to acquire iron from our blood.

Very few are so convinced of the antibacterial benefits of reduced iron stores that they advocate iron deficiency. Indeed, there is rather good evidence that our immune system, another weapon against infection, is impaired by lack of iron. It may well be that evolution has provided us with mechanisms to achieve a nearly perfect balance between iron deficiency and excess. Gross errors on either side of nutritional adequacy may be equally harmful.

The development of antibiotics has taken the terror from most bacterial infections. In contrast, many forms of cancer remain essentially incurable. For this and other reasons, a study by Richard G. Stevens and his colleagues that appeared in October of 1988 gained wide publicity; in my newspaper,

---

> *...a harmful effect of iron with respect to human cancer*
> *is plausible in light of what is known*
> *of the stimulating effects of iron*
> *on tumor growth in animals.*

---

an account of the work appeared under the headline "High Iron Levels Tied to Cancer Risk in Men." The Stevens group found that of 14,407 men and women whose iron status had been measured in the early 1970s, 445 had developed cancer by 1984 and that, at least in men, some forms of cancer were more likely in those with high levels of blood iron. Although it could not be established that blood iron was significantly related to dietary intake, a harmful effect of iron with respect to human cancer is plausible in light of what is known of the stimulating effects of iron on tumor growth in animals. Dr. Stevens' concluding sentence seems appropriate both for his report and for this chapter: "Iron supplementation for those who are not anemic may be unwise."

Summing Up

The belief that dietary iron is essential if we are to avoid anemia is entirely valid; that belief is as well much too simple. It is now clear that iron is involved in a variety of biological functions only the most obvious of which are concerned with the blood. The recognition of this great complexity has forced rejection of the notion that more iron is better. Only for pregnant women can the assumption reasonably be made that supplemental iron is a good idea. For the rest of us, a diet in which modest amounts of meat and generous amounts of vitamin C are present will insure an adequate intake of iron. Worry or fatigue or athletic activity are not reasons to take an iron pill every day. Despite my and your great desire to become independent in matters of our own health, anemia can only be diagnosed by appropriate laboratory tests. For the vast majority of us who are not anemic, the wisdom of iron supplements remains to be proven.

PART II

EXERCISE

Introduction

For most of us, exercise is not a neutral subject. Some seek to avoid it as if it were a lethal disease. Others describe their daily exercise routine as would a monk his prayers. J. V. Durnin, a Scotsman and exercise physiologist, says this about jogging: "...even sensible people pretend they enjoy it. The enjoyment is seldom obvious...I find jogging to be a particularly useless form of activity, with little apparent pleasure to be obtained from it."

I must differ with Professor Durnin. After more than 25 years, I continue to find running an almost always pleasurable experience. The fact that it seems to confer benefits other than the purely hedonistic is a bonus. (Confusion sometimes arises on the distinction between running and jogging. I am a runner; all of those few people slower than I are joggers.)

In the next two chapters, we will examine the principles that govern the training effect and I will outline a program of exercise. Emphasis in these chapters is upon aerobic conditioning and the bodily changes that inevitably accompany it. Little is said about weightlifting, strength training, or body building. The relative neglect of these latter subjects is not for want of appreciation; I engage in regular, if limited, weight training, find it quite enjoyable, and suspect that it is beneficial. On the other hand, the benefits to health of aerobic exercise are well established, whereas those of pure weight training with a minimal aerobic component are not nearly so clear.

CHAPTER 19

The Training Effect

---

*...the human machine thrives on hard use and
becomes stronger in the process....The response of
the body to use is called the training effect.*

"I consider the human body a machine." That analogy
is as useful today as when Rene Descartes first expressed it
in the 17th century. Our bodies generate energy by oxidiz-
ing food, and in the process emit carbon dioxide and other
waste products. The eyes and ears and other sensory devices
provide a central computer, the brain, with information that
is selectively stored or ignored or acted upon. The machines
of action are the muscles and, when properly guided by the
brain, they can wink an eye or run a marathon or sing a song.

Like all things mechanical, the living machinery of our
body wears out with the passage of time—the process of
aging. Then, after a few spare parts have perhaps been removed for use by others, we are recycled: "Thou art dust,
and shalt to dust return" was John Milton's elegant line. But
this notion of wearing out can be misunderstood. Unlike a
'42 Lincoln Continental that continues to escape the junkyard
because it is coddled and protected, the human machine
thrives on hard use and becomes stronger in the process. Put
your body up on blocks for the winter and you will emerge
with the spring, not bright and shiny and ready for action,
but shriveled and weak. The response of the body to use is
called the training effect.

Muscle is made up of large numbers of elongated cells or fibers. The protein in each muscle fiber is arranged in such a fashion that it can slide into itself like a hand telescope. If many fibers do this at the same time, the muscle shortens.

To understand the training effect, we need to know a little about our muscles and how they translate food into motion. Muscle is made up of large numbers of elongated cells or fibers. The protein in each muscle fiber is arranged in such a fashion that it can slide into itself like a hand telescope. If many fibers do this at the same time, the muscle shortens. The immediate source of energy for muscle contraction is a chemical found in all cells, adenosine triphosphate (ATP).

When I become aware that I have stepped into the path of an oncoming truck, my brain sends an understandably urgent message to my muscles to get me out of the way. Lengthy considerations of my vitamin status, or how tired I might be, or what I had for breakfast this morning are obviously out of order. I need immediate action. Fortunately, some ATP is present in my muscles and more can quickly be generated from muscle glucose. This is done without oxygen and that's a good thing because I don't have time to get it from my lungs to the muscles anyway. The ability to contract muscle without waiting for oxygen to arrive gives my truck-escaping behavior its name, anaerobic, an interesting word put together from the Greek words "an" (without), "aer" (air), and "bios" (life).

If, instead of a truck, it is slowly rising flood waters that threaten me and the high ground is ten miles distant, I had better respond in a different fashion. Were I to sprint off at top speed, the ATP stored in my muscles would be rapidly used up and collapse would overtake me just before the flood. If I am other than foolish, I will start at a pace I can maintain for the hour or two that it will take to cover the ten miles.

...many of us begin to exercise only when we can no longer tolerate the sight of our naked bodies and fervently wish ourselves to look different.

The large amounts of ATP needed to sustain such a prolonged effort can only be generated by providing the muscle with more nutrients and oxygen. Nutrients present no problem because my muscular engines can run on carbohydrate or fat, or even protein. For example, just one of my several pounds of stored fat is good for about 30 miles. On the other hand, the increased need for oxygen will require me to breathe more often and more deeply and my heart to beat faster. My escape from the flood demands air, hence the term, aerobic exercise. As we shall see, the consequences of anaerobic and aerobic exercise are not the same.

Exercise can make us look different. In fact, many of us begin to exercise only when we can no longer tolerate the sight of our naked bodies and fervently wish ourselves to look different. The extremes of appearance that result from exercise might be called the lifter and the runner; each is the result of a training effect. The lifter may have the well-defined muscles of a Mr. Universe contestant, but more often simply looks like he's just been blown up and may explode at any minute. The runner on the other hand has the appearance of a man coming off a death march. They look different because the weightlifter's exercise has been mostly of the brief, intense, anaerobic kind, whereas the runner has engaged in longer and necessarily less intense aerobic exercise.

Let's consider the effects of lifting weights and running on the same muscle, the 'vastus lateralis,' one of the three major muscles of the thigh. Weightlifters place a bar loaded with several hundred pounds across their shoulders, lower their bodies into a squat, and return to a standing position. One squat takes about three seconds and will be repeated

---

*The extremes of appearance that result from
exercise might be called the lifter and the runner;
each is the result of a training effect. The lifter...
often simply looks like he's just been blown up and
may explode at any minute. The runner...has the
appearance of a man coming off a death march.*

---

perhaps four times. If the weight has been chosen with care, the fourth repetition requires maximum effort and a fifth is impossible; there simply isn't enough ATP left in the muscle. Twelve seconds of anaerobic exercise doesn't seem like much, but if it is repeated three or four times a week for several weeks, remarkable changes will take place. The class of muscle cells best able to contract anaerobically, the fast twitch fibers, will increase in size and in their ability to generate ATP. The overall result will be a bigger and stronger vastus lateralis.

Anyone who doubts the reality of the anaerobic training effect need only compare the weights of football players over the past several decades. In 1964, the University of Alabama team that won the national championship had only one starting lineman who weighed more than two hundred pounds, Cecil Dowdy at 206. Less than 20 years later, in his last season of coaching, Coach Bryant's offensive line averaged 258 pounds. This difference results almost entirely from the realization by football coaches and others that anaerobic weight training can rapidly produce very large increases in muscle size and weight. (With linemen in football now routinely approaching 300 pounds in weight, some are inclined to give a part of the credit to anabolic steroids.)

In contrast with the lifter's brief, intense effort, a single stride of the runner places a much smaller demand on the vastus lateralis. However, that small demand may be repeated thousands of times without rest. Unlike the lifter, who borrows ATP from the muscle bank, the runner must pay as he goes. To do that he depends preferentially on the second major class of muscle cells, the slow twitch fibers. These are the ones best able to provide ATP from oxygen and nutrients; in other words, to contract aerobically. If the vastus lateralis is exercised by running for 30 minutes a day for a few weeks, the slow twitch fibers will increase their capacity to produce ATP. However, they will not increase in size. It's for this reason that women who take up running needn't worry about developing massive legs; aerobic exercise does not produce bulky muscles.

The training effect of aerobic exercise does not end with the slow twitch fiber. If the muscles are to use more oxygen, the whole system for delivery of oxygen must be improved. A trained muscle fiber is seen to have more capillaries, the smallest of blood vessels, surrounding it. The muscular components of the heart become stronger and more efficient so that more blood can be pumped per minute. Respiratory muscles develop the endurance to maintain deeper and more frequent breathing. The nervous system becomes better able to shunt blood from one area to another. Changes such as these separate us from all things mechanical; we respond to the stress of exercise by improving our ability to exercise; we become more fit.

Measurement of an anaerobic training effect is quite simple. How much more weight can be lifted after training than before? Matters of technique aside, one could reasonably argue that the reigning world weightlifting champions are also the best anaerobic trainers. For a typical sedentary adult, truly remarkable increases in strength can be expected from weight training.

The results of aerobic training are somewhat more difficult to quantify. Instead of just adding up weights on a bar, we need to measure the effect of exercise on the maximum amount of oxygen that a person can use. This measurement is beyond the capabilities or needs of most of us, but it is done routinely by exercise physiologists. Anyone who reads

---

A 70-year-old great-grandmother who has trained
herself to run a marathon is aerobically more fit
than many professional baseball players,
bowlers, and golfers.

---

one of the magazines devoted to running is familiar with the picture of one or another famous marathoner with a clip on his nose running on a treadmill while breathing through a cumbersome-looking mouthpiece. By changing the speed or inclination of the treadmill, the athlete can be made to give a maximum effort and, in the process, to consume oxygen as rapidly as he is able. As we become more fit, maximum oxygen uptake increases. Because the measurement of oxygen uptake is impractical for most of us, a simpler indication of aerobic effort is needed. In the next chapter, we'll see that heart rate serves this purpose very well.

Summing Up

To become fit, we need be neither exercise physiologist nor athlete. Indeed, some of our most renowned athletes, persons blessed with unusual motor skills that in another time or culture would be without value, are remarkably unfit. A 70-year-old great-grandmother who has trained herself to run a marathon is aerobically more fit than many professional baseball players, bowlers, and golfers. Our knowledge of physiology need extend no further than the distinction between aerobic and anaerobic exercise. Anaerobic exercisers, weightlifters, must be aware that, however attractive they may find their mirror images to be, their hearts and circulatory systems may be relatively unimproved after years of training. Given patience, dedication, and 90 minutes a week, aerobic fitness is within the reach of all of us.

CHAPTER 20

A Program of Exercise

"...most people do not need to see a doctor before they start since a gradual, sensible exercise program will have minimal health risks."

Is it necessary to see your doctor before beginning an exercise program? For many years if you asked that question of the medical establishment, you would get the expected answer: Yes. A statement by the American Medical Association in 1958 is typical. "All persons should be shown by medical examination to be organically sound before performing training routines (and these examinations) should be repeated periodically." This is an easy and safe bit of advice to give. Despite solid evidence of unnecessary surgery, excessive X-rays, unneeded diagnostic tests, and other indications of meddling by modern medicine, many of us accept the notion that the more often we see a doctor, the healthier we will be. However, it is also advice that is easy to ignore. How many of us are willing to devote several hours and to part with anywhere from ten to a few hundred dollars consulting with a physician before we put on our walking shoes or mount a bicycle? I don't think it's bad advice to see a doctor before beginning to exercise; it's just that for most of us it is advice that is out of touch with reality.

Fortunately for our consciences, failing to visit a physician prior to starting exercise is no longer a sin. The National Heart, Lung, and Blood Institute issued a booklet in 1981 called "Exercise and Your Heart" stating that "most people do not need to see a doctor before they start since a gradual,

> *...if the condition of your heart, joints, or other parts is less than good, or if you have no idea of what their condition is, see your doctor.*

sensible exercise program will have minimal health risks." However, "most people" doesn't include everyone, and the Heart Institute goes on to specify those who should consult a physician first. Simply put, if the condition of your heart, joints, or other parts is less than good, or if you have no idea of what their condition is, see your doctor. Of course, the phrase "your doctor" implies that you have a personal physician. If you don't, get one—whether you think yourself healthy or not. Today's medicine is too complex and too specialized for any of us to get along without an advocate. If you've had recent occasion to experience the kind of anonymous, brand-X medicine practiced in many hospital emergency rooms, no further emphasis of this point is needed.

The basis of every fitness program must be aerobic exercise. It is the aerobic training effect that best accounts for the known, presumed, and suspected benefits of exercise to be discussed in Part III of this book. Regardless of the specific form of exercise we choose, the muscles must be used in a way that will make the heart and circulatory system work long enough and hard enough to produce an aerobic training effect. The first question then is, what is hard enough? Does a leisurely stroll around the block have the same consequences as a fast run? Common sense tells us that they do not, but we need a precise way to tell the difference.

> *The basis of every fitness program must be aerobic exercise. It is the aerobic training effect that best accounts for the known, presumed, and suspected benefits of exercise...*

...heart rate is highest when oxygen uptake is greatest. It's for this reason that the beating of our hearts can be used as a guide to the intensity of our aerobic exercise.

I've mentioned previously that the maximum effort in an aerobic exercise such as running is accompanied by the greatest possible use of oxygen by the muscles; what exercise physiologists call maximum oxygen uptake. The intensity of exercise can be measured in terms of the percentage of maximum oxygen uptake that it requires—a fast run may need 90% of maximum, whereas our leisurely stroll may require less than 50%—but such measurements are completely impractical for nearly all of us.

Fortunately, we all can measure heart rate and, it turns out, heart rate is highest when oxygen uptake is greatest. It's for this reason that the beating of our hearts can be used as a guide to the intensity of our aerobic exercise. A rough estimate of your maximum possible heart rate, that which will occur with your maximum aerobic effort, is gotten by subtracting your age from 220; thus, for a 40-year old, it comes out to 180.

Our original question, what is hard enough, can now be restated as, what percentage of maximum heart rate must be maintained during exercise in order to produce an aerobic training effect? Estimates of that percentage vary and, rather than trying to state an exact figure, it is more useful to speak of a range of heart rates. For example, the National Heart, Lung, and Blood Institute gives a target zone of 60–75%; for our 40-year old, 108–135 beats per minute. This is a rather conservative range and for that reason it is a good place to start for someone who is quite unfit. As conditioning proceeds, a range of 70–85% of maximum will not be uncomfortable. For the competitive athlete whose goal is to maintain a very high level of oxygen uptake, exercise at heart rates close to 100% of maximum is not unusual.

Anyone who has run for a bus or a touchdown or an errant child knows that heart rate increases with exercise. We might then guess that the maximum rate of beating of the heart would increase as we become more fit. In fact, our formula for maximum heart rate, 220 minus age, changes very little with aerobic conditioning. Instead, the heart becomes able to pump more blood with each stroke. It is as if a bigger pump, not a faster one, had been installed. An interesting consequence of this is that fit and unfit people tend to have quite different heart rates when at rest. Put another way, as an individual becomes more fit, a decrease in resting heart rate can be expected. Most of us have learned somewhere in the distant past that a normal adult resting heart rate is about 72. That "normal" value probably represents a poorly conditioned heart; a moderate degree of aerobic training results in values of 50–60 and an elite marathoner may have a resting heart rate less than 40. (There are exceptions to the general rule that resting heart rate declines as aerobic conditioning progresses. For example, Jim Ryun, former world record holding miler, is said to have had a rather high resting heart rate.)

Question number 2: How long must we exercise to produce an aerobic training effect? The answer depends upon how hard we exercise. An activity that fails to elevate heart rate to at least 60% of maximum is unlikely to change aerobic capacity no matter what its duration. This is why many people who sit in an office and work "long, hard hours" every day can still be in very poor condition from an aerobic standpoint. On the other hand, relatively brief exercise of high intensity can quickly produce results. In one early study, for example, a very significant training effect in middle-aged, sedentary men was achieved with only 36 minutes of exercise a week, but it was all at about 85% of maximum heart rate.

My suggestion is that everyone work toward an initial goal of three 30-minute sessions per week at 70–80% of maximum heart rate. Once achieved, this combination of duration and intensity is likely to produce all the known benefits of aerobic training. Ultimately, each of us will find our own most agreeable compromise between how long and how hard we exercise.

So long as heart rate is increased sufficiently and maintained at an elevated rate long enough, it matters not at all what particular form of exercise we choose. However, because running is one of the simplest and most efficient means of training and because it has been so well-studied in recent years, I will use it to illustrate a beginning program. In 1967, a group headed by Dr. Waldo Harris at the University of Oregon published an article in the *Journal of the American Medical Association* with the simple title "Jogging." They defined jogging as "walking and running alternately at a slow to moderate pace." Men ranging in age from 30 to 66 exercised three times per week on a quarter mile track. They began with a total distance of one mile, equally divided between walking and running. Over a period of 12 weeks they progressed to about 2-1/2 miles, mostly running. By today's standards, the demands of the program might be considered inadequate, but the essential feature, a gradual increase in intensity and duration of exercise, remains to guide us. (A coauthor of "Jogging" was William Bowerman, MA, then the track coach at the University of Oregon, who was later a founder of the Nike Shoe Company.)

Once you have accepted the discipline of regular exercise and have begun to feel its value, you no longer will have need of preachers like me. Until that time comes, a rather rigidly defined program can lend some support. These are my rules for starting:

(1) Choose the quarter of the year when you are most likely to be able to devote 30 minutes on each of three days per week to your renaissance. The first month is the most important. For example, it might not be a good idea to start on December 15th; in my part of the country you will have people telling you it's too cold to exercise outside (it rarely is) and no matter where you are, holidays can be a distraction.

(2) Commit a specific time of day to exercise. It matters not at all when it is, but it should be decided in advance. After a few months, when exercise is established as a part of your routine, you can be less rigid about this. In the early 60s, from sheer embarrassment, I ran in the dark of night. Now, out of the closet, I run whenever it suits me and my other activities.

(3) Decide where your jogging is to take place; try to shun traffic, dogs, and other hazards and distractions. Any fairly smooth surface will do, but avoid concrete if you can.

(4) Be gentle. Set your own pace. The results of years or decades of neglect will not be reversed in a week. Some will have trouble walking for 30 minutes; a few will be able to run that long right from the start. Gradual progress is what we're after. Use your heart rate to measure effort. (Most of us feel a little silly taking our pulse in public. Be assured that once you get to know your body a little better it will become unnecessary to measure heart rate.)

(5) Keep a record of your activities. This can be as elaborate as you like, but the minimum is the date and duration of your exercise. Use one week as the basic unit of time and try always to exercise three times in each week. To minimize muscle soreness, take a day off after each day of exercise rather than lumping the three days together. If two of your three days must be on the weekend, make Sunday's workout a little easier than Saturday's.

(6) Focus on aerobic exercise for three months. Don't diet, don't stop smoking, don't do anything that will distract, distress, or discourage you. As you become more fit, you may find that cigarets and heavy foods are less attractive to you. However, specific weight loss and other major changes are best put off until aerobic conditioning is well established. (In Chapters 23 and 25, I'll tell you in more detail why I suggest that exercise should come before dieting.)

Some people make a big thing of warming up and cooling down. My opinion is that it is more a matter of personal

---

...the basic equipment needed for a scientifically valid program of aerobic running is a comfortable pair of shoes, clothing adequate for decency, and a time piece. That's it.

---

preference than physiological necessity. If you want to do a few gentle stretching exercises before or after running, go ahead, but don't count it as part of your 30 minutes. My practice is to use the first five or ten minutes of a run as a warmup. The fact is that there has been very little decent research done on the merits of exercises for flexibility, but I suspect that it is very much an individual matter; if you like to stretch or think that it does you some good, do it. Don't confuse it with aerobic exercise.

Sometimes it is hard to tell the difference between a running magazine and a fashion magazine. Each have their pieces of advice on diets and health and beauty, and each is loaded with advertisements for shoes and clothing. It's easy to forget that the basic equipment needed for a scientifically valid program of aerobic running is a comfortable pair of shoes, clothing adequate for decency, and a time piece. That's it.

Although I ran for a long time in a pair of work boots from Sears, I have to admit that most of us are better off in shoes specifically designed for running. Until you develop your own likes and dislikes, you won't go wrong with a basic pair from Adidas or Nike or New Balance or Reebok. A training shoe in the middle of their price range will cost $30–50 1990 dollars. The best place to get them is a store that caters to the needs of runners and where most of the sales people are a bit gaunt. This means that they probably are serious runners and, as such, are founts of information and sometimes wisdom about shoes. Your timepiece needn't be elaborate, but a digital wrist watch/stop watch is very useful and can be had for less than $25.

Three months have passed and your records indicate that the goal of three 30-minute sessions per week has been met and maintained. Now is the time to make adjustments in your program. Perhaps you prefer another form of exercise such as rowing, cycling, or swimming; they're all fine so long as the principles of aerobic conditioning are not forgotten. For example, you will want to recheck your heart rate during exercise to make sure that you are reaching the target zone. If you continue to run, you may wish to do it for distance rather than time. Instead of 30 minutes a day, switch to a distance that takes about that long. Experiment. If you feel like running every day, do it. If longer distances appeal to you, run them. If you want competition, enter a two or three mile fun run. You may discover a talent you never knew was there. Finally, start to consider other changes you may wish to make in your life. For example, if you are a smoker or you are overweight, you need to look no further for your next project.

Summing Up

Video tapes, health clubs, designer outfits to sweat in, and other forms of the commercialization of fitness sometimes obscure the simple beauty of the principles of conditioning. If your body is basically healthy, and you may wish to get an expert's opinion on that, nothing stands between you and an aerobic training effect but you. Using heart rate as a guide and 30-minute sessions as a goal, you will reach a basic level of aerobic fitness in a remarkably short period of time. Sometime during that period the pleasure will begin; sophisticates will insist that you have increased the level of endorphins in your brain. Whatever its origins, the pleasure may take forms as varied as the individuals who experience it. Mine might best be called tranquility. Meet me on the road and we'll talk about yours.

PART III

INTERACTIONS

Introduction

In the Preface to this book, I said that we would strive for total perspective. It is in these five concluding chapters that the attempt will be most evident. Although it is difficult to imagine a disease in which nutrition and exercise do not play a proven or potential role, I have focused upon three major afflictions: cancer, heart disease, and osteoporosis. For all three, nutrition seems clearly to be of significance, whereas for osteoporosis and heart disease, exercise is certainly of importance as well.

The final chapter is concerned with ideal weight and deviations from it. In some ways, it is the most complex subject of all. No arguments arise over whether osteoporosis, cancer, and atherosclerosis are diseases. The same cannot be said for excess fat; indeed, we have trouble defining the boundaries of "excess." It seems proper to call extreme overweight a disease and to give it a proper medical name, obesity. But for the majority of us, overweight is most certainly a nondisease whose medicalization is without merit.

CHAPTER 21

Cancer

Can 90% Be Prevented?

...our present understanding of the diseases
called cancer is woefully inadequate,
much of the advice and treatment of today will
inevitably be shown to be bad advice and bad treatment.

Cancer is a disease that has afflicted humans for as long as we can remember. The ancient Greeks called it "karkinoma," which became the Latin word "cancer" with the dual meaning of a crab or of an ulcer that moves crablike from one spot to another. In addition to this tendency to spread from the site of origin to new places in the body, a phenomenon now called metastasis, cancer is characterized by uncontrolled growth of cells. Cancer may kill us in a variety of ways: Sometimes the liver or the brain or another essential organ is invaded and its function disrupted; sometimes there is general wasting of the body and its defenses to the point that overwhelming infection occurs.

We most often use the singular form in speaking of cancer; we shouldn't. Cancer is not one disease but a whole family of diseases, perhaps 100 or more in number, each with its own peculiarities of origin and course. For this reason, the potential means to avoid "cancer" are varied and often seem unrelated. Indeed we shall later consider the possibility that measures taken to prevent one kind of cancer may increase the probability of another. We must ever keep in mind that

---

...the war [on cancer]...is not being won...
mortality from cancer has remained unchanged
for the past 30 years. [Some] argue that our
emphasis must be shifted from cure to prevention.

---

our present understanding of the diseases called cancer is woefully inadequate; much of the advice and treatment of today will inevitably be shown to be bad advice and bad treatment. Finally, despite the pessimistic sound of the preceding, there is overwhelming evidence that some forms of cancer can be prevented entirely or at least postponed until some more agreeable cause of death overtakes us.

We have come to expect that modern medical science, given enough time and money, can find a cure for every disease. That expectation justifies our "War on Cancer" and the billions of dollars lavished upon cancer investigators. But critics such as John C. Bailar, III, of the Harvard School of Public Health have suggested that the war, focusing as it does on the cure of disease once established, is not being won; that mortality from cancer has remained unchanged for the past 30 years. He and others argue that our emphasis must be shifted from cure to prevention.

There is no question that prevention is far preferable to cure; one need only consider the enormous cost in dollars and misery and failure of today's treatments for cancer. But, given cancer's roots in antiquity, what stretch of the imagination leads us to believe that it can be prevented? The answer lies in a simple observation: The incidence of specific cancers is quite different from place to place. For example, American women are five times as likely to die of breast cancer as are Japanese women, whereas stomach cancer develops in Japanese men ten times as often as in their American counterparts. Does this mean that Japanese women have a genetically determined resistance to breast cancer while

> *...what stretch of the imagination leads us to believe that it can be prevented? The answer lies in a simple observation: The incidence of specific cancers is quite different from place to place.*

Japanese men have an inherent suceptiblility to stomach cancer? It does not: Move the Japanese to America and in two generations they will develop cancer as Americans do. International comparisons of cancer incidence suggest that external factors markedly alter the probability of occurrence of many forms of the disease. But there has been much confusion regarding the nature of these "external factors." Influenced in large measure by a flood of information and misinformation on the subject, many people have decided that "everything causes cancer" so there's no point in worrying about anything. Others have come to believe, wishfully perhaps, that a little more fiber, or a vitamin supplement, or avoiding food additives will reduce the risk of cancer to the vanishing point. The truth, as best we know it today, lies somewhere between these two extremes.

In 1952, John Higginson, an Irish physician and scientist, concluded, on the basis of his comparisons of Blacks living in Africa and in the United States, that two-thirds of all cancer is environmentally determined and thus preventable. Later, as the founding director of the World Health Organization's International Agency for Research on Cancer, Dr. Higginson did much to educate the world about the preventability of cancer. Unfortunately, his message was often misunderstood and sometimes willfully distorted. The problem centers on the diverse meanings given such words as "environment," "behavior," and, one of today's favorites, "lifestyle."

Higginson defined the word environment as "what surrounds people and impinges on them"; not far from my dictionary's definition as "the total circumstances surrounding us." Environment, as Higginson intended, includes virtually all factors in human life other than genetic endowment; and even genes can be altered by environmental factors. Professional and amateur "environmentalists" have tended to focus upon clean air and pure water—important items, but hardly the whole story. For the Higginson environment, we must add sunlight, cosmic rays, radon from the earth, sexual practices, age of menstruation, number of children, lifetime habits of exercise, history of viral and bacterial infection, every element of the diet, medical treatment, tobacco, alcohol, and, no doubt, a whole lot of things not yet thought of. Enlarged in this way, environment is reasonably implicated in the majority of cancers.

...many people have decided that "everything causes cancer" so there's no point in worrying about anything. Others...believe, wishfully perhaps, that a little more fiber, or a vitamin supplement, or avoiding food additives will reduce the risk of cancer to the vanishing point. The truth, as best we know it today, lies somewhere between these two extremes.

Few of us have any real interest in all of the factors that might have some bearing on the causation of cancer; right now the whole area is simply too messy and too uncertain. What we need is a point of reference; something of known cancer-inducing efficiency against which to measure everything else. Tobacco provides that point of reference.

In 1912, Irving Adler apologized to his readers for writing at such great length about a disease so uncommon as lung cancer. By 1986, lung cancer was the leading cause of death from cancer in both men and women in the United States. In that year, 134,000 Americans died of the disease; fully one-third of all cancer deaths among men. Now add in the contribution of tobacco to cancer of the mouth, tongue, larynx, esophagus, and bladder, to heart disease, and to a variety of noncancerous diseases of the lungs, and we do indeed have a towering point of reference, one unapproached by any other environmental factor. (Compare AIDS, acquired immune deficiency syndrome, a disease of astonishing lethality that has set terror in the minds of many. A total of 60,000 persons died of AIDS in the first eight years following its appearance in this country in June of 1981. That many more will die is as certain as it is tragic. That AIDS will approach the cumulative toxicity of tobacco is unlikely.)

The *British Medical Journal* for December 13, 1952 included an article by Richard Doll and A. Bradford Hill entitled "A Study of the Etiology of Carcinoma of the Lung." Based upon interviews with 1488 patients with lung cancer and an equal number of comparable persons without the disease, they concluded that "the association between smoking and carcinoma of the lung is real." By 1957, the Medical Research Council of Great Britain went further: "the relationship (between smoking and cancer) is one of direct cause and effect." The MRC was at that time aware of a study that would appear in the March 15, 1958 issue of the *Journal of the American Medical Association*. In it, Cuyler Hammond and Daniel Horn of the American Cancer Society reported their conclusion, based on a study of 187,783 American men, that among regular smokers of cigarets, perhaps four deaths in 10 might be caused by that habit.

By 1964, even the Surgeon General of the United States was ready to act: "The magnitude of the effect of cigaret smoking (on lung cancer) far outweighs all other factors." In 1979: "Cigaret smoking is the single most important preventable environmental factor contributing to illness, disability, and death in the United States." The case against tobacco has grown stronger in each succeeding year. The best estimate for 1986 was that 87% of all lung cancer deaths in this country would have been avoided if no cigarets had been smoked;

that's 117,000 lives! Against this backdrop, the pictures of happy, healthy, beautiful young people in advertisements for cigarets are the truest obscenity in the media today.

Let us now turn to cancer prevention for nonsmokers. Here the issue is considerably more complex; we must begin to consider all those nontobacco factors in the Higginson environment. Estimates of risk are quite varied; those given below were provided by Richard Peto of Oxford University in 1985; no remarkable changes have occurred since then. I have chosen Peto's figures because he, together with his collaborator of many years, Sir Richard Doll, is the world's most authoritative source of such information.

Peto provides us with two estimates. The first he calls the "present imperfect": reliably established ways of avoiding the onset of life-threatening cancer in the United States and other developed countries. The imperfection of the list lies in the fact that today's knowledge, applied with total efficiency, could be expected to prevent less than 40% of all cancer deaths. Furthermore, fully three-quarters of the known-to-be-preventable cancer is caused by tobacco smoke. That seems to leave very little for the nonsmoker to do. Factors we hear much about—the medical use of hormones and radiation, excessive exposure to sunlight, pollution of air, water, and food—can each account for less than 1% of cancer deaths.

Before moving on to more speculative matters, I must point out that population statistics such as those in Peto's present-imperfect list can be seriously misleading. For example, Richard Doll described a group of 19 men exposed to 2-naphthylamine, a solvent used in the dye industry. Eighteen died of bladder cancer and the last avoided that fate only by being killed in an accident. Thus, a factor that is insignificant in the population as a whole may be nearly 100% efficient in causing cancer in those exposed to it. It is for this reason that we must continue to monitor the environment for newly introduced hazards and to protect workers and others from

> *In the "future perfect" we would know with certainty*
> *the contribution to various cancers of...alcohol,*
> *sexual behavior, hormones, reproductive variables,*
> *occupation, pollution, drugs, and other medical*
> *treatments, sunlight and cosmic radiations, infection,*
> *and diet. Furthermore...we would not only know the*
> *risks presented by these factors, but we would also be*
> *able to exert near perfect control over them.*

known hazards. Nonetheless, in terms of today's knowledge, tobacco is far and away the most important and well-documented cancer-causing factor in the population as a whole.

If lung cancer is the only form of the disease that has increased dramatically in this century, what leads Peto and Doll and Higginson and others to suggest that 80–90 percent of all cancer might be avoided? The answer is found in Peto's second list of estimates; he calls it the "future perfect." In the future perfect we would know with certainty the contribution to various cancers of such diverse factors as alcohol, sexual behavior, hormones, reproductive variables, occupation, pollution, drugs, and other medical treatments, sunlight and cosmic radiations, infection, and diet. Furthermore, in the future perfect, we would not only know the risks presented by these factors, but we would also be able to exert near perfect control over them.

With the example of smoking and lung cancer before us, it's not hard to see why some are not optimistic about ever reaching Peto's future perfect. We know with a certainty rare in medical science that the use of tobacco contributes to lung cancer and a variety of other diseases, yet we continue to permit the seduction of young people by billions of dollars worth of advertising and we continue to subsidize the growing of the plant by American farmers.

Now consider a more subtle factor from the future perfect. There is some evidence that the onset of menses at an
early age is a risk factor for subsequent breast cancer. The
age at which menstruation begins is in turn related to development in general; big, tall girls menstruate earlier than small,
thin ones. Development in general is clearly influenced by
diet. If, in the future perfect, it could be proven that optimal
growth from birth to puberty slightly increases the incidence
of breast cancer 30 or 40 or 50 years later, would we embark
on a program of intentional underfeeding of our daughters?
I doubt that we would; for good reasons and bad.

Enough of subtleties and unknowns and perhaps
unknowables; they tend to give one a headache. Let's now
consider fat and fiber and fruits and vitamins and minerals
and all those other features of our diet that, at least as told in
the popular media, surely will prevent cancer. Peto's best
guess for diet-related cancer is 35%, but he acknowledges
acceptable estimates from 10 to 70%.

Just as the vague notion that smoking is associated with
lung cancer had been around for 200 years before the issue
was proven beyond doubt, the idea that cancer might be
related to what we eat is not new; it has bobbed in medicine's
consciousness for several thousand years. But cancer is such
a common disease with so many possible causes that it was
not until the late 1960s that adequate data were available to
permit beginning attempts to correlate specific cancers with
elements of diet. Richard Doll would again play a major role.

In the April 1975 issue of the *International Journal of
Cancer* was an article entitled "Environmental factors and
cancer incidence and mortality in different countries with
special reference to dietary practices." Its authors were Bruce
Armstrong and Doll, who at the time was Regius Professor
of Medicine, Radcliffe Infirmary, Oxford University. The article is long and complex and filled with statistical jargon,
but its conclusions are admirably simple and modest and
provocative.

Armstrong and Doll correlated incidence rates for 27 cancers in 23 countries and mortality rates for 14 cancers in 32 countries with a wide range of dietary factors. In their words "the strongest points to emerge from these analyses are the suggestions of associations between cancers of the colon, rectum and breast and dietary variables—particularly meat (or animal protein) and total fat consumption...it is clear that these and other correlations should be taken only as suggestions for further research and not as evidence of causation or as bases for preventive action...(however), it is possible that diet may have an effect upon many cancers...the subject warrants more attention...." And more attention it got.

In Chapter 24, I define a nutrition activist as one who favors the early translation of scientific findings into advice for the general public. Of course the pitfall in moving too quickly from laboratory and clinic to the mass media is that subsequent investigations may require modification or even downright repudiation of earlier advice. Although professional scientists are comfortable with uncertainty, their fellow citizens as well as their elected representatives take a dim view of regular revision of what had been presented as dogma. Nutrition activism, and its perils, are well illustrated by the issue of diet and cancer.

In June 1980, the Division of Cancer Cause and Prevention of the National Cancer Institute commissioned a study of diet and cancer by the National Academy of Sciences. On June 16, 1982, the 14-member committee issued a 500-page report simply titled "Diet, Nutrition, and Cancer."

The conclusions reached by the committee may be summarized as follows:

(1) Fat, both saturated and unsaturated, should represent 30% of the caloric intake. (The present average in the United States is about 40%).
(2) The intake of fruits, vegetables, and whole grain cereals should be increased.

(3) Alcoholic beverages should be used only in moderation.

(4) Consumption of foods preserved by salt-curing, salt-pickling, and smoking should be minimized.

(5) Present evidence does not permit recommendations with respect to total caloric intake, protein, fiber, carbohydrates, cholesterol, most minerals (including selenium), vitamin E, and the B vitamins.

(6) Mutagens in foods, food additives, and environmental contaminants do not contribute markedly to the risk of cancer in the United States.

On the one hand the committee was cautious: "...cancers of most major sites are influenced by dietary patterns but the data are not sufficient to quantitate the contribution of diet to the overall cancer risk or to determine the percent of reduction in risk that might be achieved by dietary modifications...it is not now possible and may never be possible to specify a diet that would protect everyone against all forms of cancer." In the words of the committee chairman, Clifford Grobstein, professor of biology and public policy at the University of California at San Diego, "Certainly we have no ideal cancer-preventing diet to announce." On the other hand there could be no doubt that the report, with the approval of the National Cancer Institute, was telling people that changes in diet would reduce the risk of cancer; this had not been done before.

Not surprisingly, the American Meat Institute and the National Meat Association, who see any anti-fat advice as being anti-meat, characterized the report as "a simplistic approach to a very complex question" and as "misleading advice which does no service to the public." John R. Block, Secretary of Agriculture, was "not so sure government should get into telling people what they should or shouldn't eat." Critics pointed out that just two years earlier, the Food and Nutrition Board of the National Academy of Sciences in a publication called "Toward Healthful Diets" had noted that

> *...the [National Academy of Science committee's]*
> *report, with the approval of the National Cancer*
> *Institute, was telling people that changes in diet would*
> *reduce the risk of cancer; this had not been done before.*

there was no basis for recommendations to modify the proportion of fat consumed by Americans. But, taken as written, few objected strongly to the report; consumption of less fat and alcohol and eating more fruits, grains, and vegetables makes sense for reasons quite unrelated to cancer. In fact, the diet suggested to reduce cancer risk is little different from that presently favored to avoid heart disease or to minimize the effects of diabetes.

The report was not always "taken as written." Predictably, every quack in the country who had ever advocated a change in diet as a means to prevent disease saw "Diet, Nutrition and Cancer" as vindication of "natural healing." (I must confess ignorance of what unnatural healing might be.) The prize for audacity perhaps should go to Mishio Kushi. Just a year after the appearance of the committee report, Kushi published a book called *The Cancer Prevention Diet*. This is what he has to say about lung cancer. "To prevent and *heal* lung cancer (my emphasis), first, all extreme foods from the yang category are to be avoided or minimized including meat, poultry, eggs, dairy products, and seafood as well as baked flour products. It is also necessary to avoid foods and beverages from the yin category, including sugar and all other sweets, fruits and juices, spices and stimulants, alcohol and drugs, as well as all artificial, chemicalized (*sic*), and refined food." Nothing in the report justifies such statements.

Books such as those by Kushi are easily recognized by most as nonsense. Unfortunately, more subtle exaggerations of the committee report appeared as well; some were written by individuals with bona fide medical training. For example,

> *The simple truth is that we don't today know
> what specific changes to make in the American diet
> in order to minimize the incidence of cancer.*

"The Doctor's Anti-Breast Cancer Diet" by Sherwood L. Gorbach, MD, and his colleagues expressed their belief that "you may reduce your risk of (breast cancer) by as much as 50 percent." The *Harvard Medical School Health Letter* stated with confidence that "the association of breast cancer with a diet high in animal fat and low in fiber is real."

Even a few members of the committee itself seemed dazzled by the spotlight turned upon them. "Study Finds Diet is Key Factor in Cancer Risk" was the title of an item sent to the newspapers of the country by the Associated Press. The basis of the article was a Chicago Tribune interview with T. Colin Campbell of Cornell University. In the interview, Dr. Campbell said that "diet, which is associated with 70 percent of all cancer, may be the most important risk factor for cancer." The implication of such statements is that if people would just watch what they eat, 70% of all cancer could be prevented. The simple truth is that we don't today know what specific changes to make in the American diet in order to minimize the incidence of cancer. The committee's relatively conservative recommendations were quickly submerged by a wave of wishful thinking.

Aside from intentional or unintentional misinterpretaton of the report, why should some eminently qualified scientists label the committees advice as "premature?" Let's consider fat. The committee recommends that we consume less total fat as a means to reduce the risk of cancer of the breast, colon, and prostate. Indeed, "the committee concluded that of all the dietary components it studied, the combined epidemiological and experimental evidence is most suggestive for a causal relationship between fat intake and the occurrence of cancer."

The epidemiological evidence to which the committee refers began to accumulate in the late 1960s. There are strong and undisputed international correlations of per capita consumption of meat and fat with the number of deaths from cancer of the breast, colon, and rectum. For example, the rate of death from breast cancer for Japanese women has been among the lowest in the world. However, American women of Japanese descent experience breast cancer at a rate nearly that of American Caucasians. One of the obvious differences between traditional Japanese and American diets is the higher proportion in the latter of calories derived from fat. Data such as these were cited by Armstrong and Doll in their landmark study of 1975. It was upon such data that the committee focused in making their recommendation to reduce dietary fat to about 30% of total caloric intake.

If a high-fat diet causes cancer or increases the risk of cancer, one should be able to demonstrate that people with cancer have consumed more fat than people without cancer. In fact, three large case-control studies in which women with breast cancer were compared with closely matched women without the disease have failed to show any consistent association with fat. These and other similar data explain why the National Cancer Center Research Institute of Japan, where our story of fat and breast cancer began, makes no recommendation to the Japanese people for changes in dietary fat in its "Twelve Tips for Preventing Cancer."

What of the "experimental evidence" to which the committee refers in recommending less fat in the diet as a means to avoid cancer? In laboratory animals, a high-fat diet increases the risk of chemically induced cancers. Although extrapolation from animals to humans is a risky business, the laboratory evidence would appear to support the committee's recommendation. However, an important distinction between types of fat must be made. It appears in such laboratory experiments that the culprit is polyunsaturated fats (PUFA) rather than fat in general. These are the same

---

*Victor Herbert, physician, lawyer, and professor
of medicine...has said that advice should come
from evidence, not inference.*

---

polyunsaturated fats that those concerned with heart disease tell us should be increased in the diet. In the absence of precise information regarding the benefits to be derived from PUFA with respect to heart disease and the adverse effects of PUFA with respect to cancer, we can make no rational decision. In fact, the case for PUFA as a means to prevent heart disease (Chapter 22) is much stronger than the case against PUFA because of concerns about cancer. Read on. Advisory committees, like nature, abhor a vacuum. It is virtually inconceivable that a committee meeting for two years at a cost to the American taxpayer in excess of a million dollars would issue no advice. It is totally inconceivable when we recall that the committee in question was assembled by the nutrition activists of the Division of Cancer Cause and Prevention of the National Cancer Institute. Ridicule would be the only reward for the humble conclusion "We need more information."

Victor Herbert, physician, lawyer, and professor of medicine at Mt. Sinai School of Medicine, has said that advice should come from evidence, not inference. That the "enthusiasts," as Dr. Herbert calls them, of the Committee on Diet, Nutrition and Cancer relied on inference is without question. On the specific issue of dietary fat and breast cancer, it now is clear in retrospect that more evidence was needed.

On New Year's day 1987, there appeared in the *New England Journal of Medicine* an article entitled "Dietary Fat and the Risk of Breast Cancer." The senior author was Walter C. Willett, MD, of the Harvard Medical School. Willett and his colleagues reported a study begun in 1980 of 89,538 registered nurses in the United States. The answers to a detailed dietary questionnaire provided a good estimate of each nurse's consumption of a number of nutrients including fat, both saturated and unsaturated.

The Willett study was prospective, i.e., none of the women had a history of cancer. Instead, the health of each nurse was monitored for four years after entry into the study. In this way the incidence of breast cancer during that period could be correlated with components of the diet in a straightforward and unbiased manner.

Of the nearly 90,000 nurses, 601 developed breast cancer during the four-year period of followup. Nurses who ate the most fat, about 44% of total calories, differed not at all in terms of likelihood of breast cancer from those who ate the least fat, about 32% of total calories. The authors' conclusions were unequivocal: "...we found no evidence that total fat intake or consumption of specific types of fat among women was positively associated with the risk of breast cancer...Our findings suggest...that a reduction in total fat intake of approximately 25% by women is unlikely to cause a substantial decrease in the incidence of breast cancer."

What then are we to make of the oft-cited correlations between dietary fat intake in a country and the incidence of breast cancer? A probable explanation is that fat, like many other features of the Higginson environment, simply co-varies with some truly causative factor. For example, international correlations between incidence of breast cancer and gross national product are about as good as those with fat. No one has yet suggested that we can prevent breast cancer by reducing the GNP. In 1983, Susan Helmrich and her colleagues of the Boston University School of Medicine reported a case-control study of 1185 women with breast cancer. They found the incidence to be higher in those women with 12 or more years of education. Will women experience less breast cancer if we stop educating them? Obviously not.

In view of the strong, some would say unwarranted, stand against fat taken by the Committee on Diet, Nutrition and Cancer, it came as a surprise to many that fiber was not given equal billing. Our own consideration of fiber in Chapter 3 led us conclude with Burkitt that the diets of most of us would benefit for a variety of reasons by the inclusion of more vegetable and fruit fiber. With regard to cancer, however, the committee stated that there is "no conclusive evidence to indicate that dietary fiber...exerts a protective effect." "Oh, never mind," say the nutrition activists, "if we don't like a committee conclusion, we'll just ignore it and substitute one that strikes our fancy."

Peter Greenwald, MD, is director of the Division of Cancer Cause and Prevention of the National Cancer Institute. It was this group that funded the Committee on Diet, Nutrition and Cancer and thus had a significant interest in the application of the committee's findings. In May of 1984, Dr. Greenwald met with officials of the Kellogg Co. regarding the company's products in relation to cancer prevention. As a result of those discussions, Kellogg cereal boxes were given a new message. Under the heading "Preventive Health Tips From the National Cancer Institute," the breakfast reader is told among other things that "eating the right foods may reduce your risk of cancer...eat high fiber foods."

In approving advertisements such as those by the Kellogg Co., Dr. Greenwald and the Division of Cancer Cause and Prevention have done more than just ignore the laboriously (and expensively) assembled recommendations of their own committee. They have as well come into direct conflict with the Food and Drug Administration (FDA).

The FDA's mandate regarding foods and advertising is found in the Federal Food, Drug, and Cosmetic Act: "A food labeled under the provisions of this section shall be deemed to be misbranded...if its labeling represents, suggests or implies that the food because of the presence or absence of certain dietary properties is adequate or effective in the prevention, cure, mitigation, or treatment of any disease or symptom." In a single pass, the Division of Cancer Cause and Prevention of the National Cancer Institute ignored the advice of a committee that it funded (with our money), joined the hucksters at your local health food store, and, at least in the view of this nonlawyer, violated the law.

Ethyl alcohol, usually called simply alcohol, is not only the intoxicating principle in all alcoholic beverages, but is also a food. As a food, it was part of the dietary questionnaire completed by the 89,538 nurses in Walter Willett's study of dietary factors in breast cancer. As we've already seen, no correlation with dietary fat was found. In contrast, Willett and his colleagues reported in May of 1987 that women who had one or more drinks per day were at 30–60% increased risk for breast cancer.

Substantial evidence for a link between alcohol and breast cancer was first presented in 1984 and most subsequent studies, including Willett's, had supported the relationship. What I find most interesting is not the epidemiology and the science involved, but the reponse to Willett's report by those who would advise the public regarding nutrition and cancer.

From what we've seen earlier of Peter Greenwald, advocate of low fat and high fiber to prevent cancer, we might expect a campaign against alcohol by the Division of Cancer Cause and Prevention. As well, we might expect "conservatives" such as L. Saxon Graham, PhD, Chairman of the Department of Epidemiology and Preventive Medicine at the State University of New York, Buffalo to point out weaknesses in the studies and the great difficulty in concluding cause and effect from simple correlations in epidemiological data. This is what actually happened. Greenwald was quoted in *Time* magazine: "We don't have the information to be making a public recommendation at this point." Graham, in editorial comments in the *New England Journal of Medicine*: "One might recommend, then, that women at especially high risk for breast cancer, such as those who are obese, who have had few children, who were first pregnant when they were older than 25, or whose mothers had breast cancer, should curtail their alcohol ingestion...it behooves us to make recommendations for prevention even while continuing scientific inquiry." Were it a horse race, my money would be on Professor Graham.

Now let's go back to polyunsaturated fat (PUFA) for a moment. It is representative of those dietary components that may be good for one thing and bad for another. The evidence is strong that a diet high in PUFA reduces the risk of coronary heart disease, yet there is as well the suspicion that PUFA may contribute to various cancers. In a like fashion, there is reason to believe that things we do to prevent one kind of cancer may increase the probability of another form of the disease. This risk stems from the fact already noted that cancer is a single name for a variety of conditions. Beta-carotene will illustrate the point.

In Chapter 5 we considered beta-carotene and its role as a precursor of vitamin A. Epidemiological studies have repeatedly provided evidence of a reduced risk of various forms of cancer in those persons who regularly consume fruits and vegetables rich in carotene. The apparent protective effect is not shared by vitamin A itself. This suggests that beta-carotene has important functions, including possible anticancer effects, in addition to serving as a source of vitamin A.

The epidemiological evidence for a protective effect of carotene is sufficiently strong that in 1982 the National Cancer Institute funded the Physicians' Health Study; in Chapter 2, I told you about the aspirin component of the investigation. More than 22,000 male physicians in this country are enrolled. Half of the group take 50 mg of beta-carotene every other day and the other half a placebo. None are aware of the group to which they have been assigned. Unlike the results with aspirin, it is still too early to draw even preliminary conclusions about the value of these beta-carotene supplements in preventing or delaying the development of cancer.

Well, if its good enough for American physicians, why shouldn't we all be taking beta-carotene pills? The answer comes in two parts. First, until the study is much further along, no one knows whether any benefit will be derived from

*All of us must live our lives today, not in some
unspecifiable future time when every knot of
ignorance regarding cancer prevention and cause is untied.*

the extra carotene. We must remind ourselves that the epidemiological studies suggesting a protective effect dealt with foods that contain beta-carotene; no one has ever shown that it is the beta-carotene rather than some other component of these foods that is important. Second, and more important, until the study is much further along, no one knows whether harm will be done.

Harm from beta-carotene? What anti-nature radical would suggest that? One who would, and he is neither anti-nature nor a radical, is the aforementioned Saxon Graham. The epidemiological studies by Professor Graham and his group have found, as have others, that those who regularly eat carotene-containing fruits and vegetables have a decreased risk of cancer of the lung, bladder, mouth, larynx, breast, cervix, and ovary. But, a high intake of carotene was associated with an *increase* in the risk of prostate cancer. Dr. Graham: "Although the contradiction is disturbing, it may not be a contradiction at all. I see no reason why a given anti-carcinogen necessarily must be an anticarcinogen for all types of cancer." Once again, cancer is many diseases.

We've now spent rather a long time learning why many of the things that many of us have believed about cancer cause and prevention are unlikely to be true. Saxon Graham put it this way: "...for most nutrients of which I am aware, too little investigation has been reported to serve as a basis for recommendation as to a prudent diet. For the most part, recommendations are based upon findings which are contradicted by other findings."

All of us must live our lives today, not in some unspecifiable future time when every knot of ignorance regarding cancer prevention and cause is untied. What is the best advice to be had? What should be done by those who want to do something? First of all, we must acknowledge to ourselves that on any given day we are merely approximating the "best" course of action; approximating because the best isn't known today and may not ever be known. I like to think of this as living with a few superstitions.

Consider my breakfast superstition. On most every morning of the year my breakfast consists of a large glass of orange juice and a bowl of high-fiber cereal with skim milk. Do I eat this "healthy" breakfast because I don't like sausage, bacon, pancakes, eggs, and gooey sweet things and all the other delights that nature and a good chef can tempt us with? Not at all. I do it because of the evidence that suggests, but hardly proves, that a low-fat, high-fiber, low-calorie, vitamin-rich diet may be good for me in a variety of ways, many of which we've already considered in this book. My superstition is less than faith, so that I occasionally eat, without guilt, all the "unhealthy" things. Indeed, I suspect that I enjoy them all the more for not having eaten them every day.

How do my superstitions differ from those of Mishio Kushi or Marilyn Diamond or Stuart Berger or Durk Pearson? There are two main differences. The first is that I acknowledge my base of information to be incomplete; indeed, that it will ever be incomplete. My superstitions are therefore ever subject to change. But, in fact, the rate of change is quite slow; more a glacier than an avalanche. The second is that my superstitions are based on the whole of what I know, or think I know, about health and disease and the factors that contribute to each. Entire books have been devoted to a single nutrient and a single disease. But life doesn't come in pieces; we get it whole or not at all.

The most reliable and, at the same time, the least acceptable way to avoid cancer is to die at an early age. In the United States, about 50 percent of all cancers are diagnosed in those over age 70. Pessimists have suggested that cancer is as inevitable as death; live long enough and the disease will appear in one form or another. Whether the age-related increase in cancer results simply from causal factors in the Higginson environment or is more directly tied to some aspect of aging is not known.

Having rejected early death as a way to avoid cancer, we may turn to smoking and other forms of tobacco use. Tobacco is far and away the most important preventable cause of cancer, indeed of premature death in general, in this and other developed countries. If there is one thing you would do for your children, raise them in a smoke-free home; from an early age teach them that smoking is a mark either of ignorance or foolishness. With our great and justifiable concern about "drug abuse," it is important to realize that the total removal of cocaine, marijuana, LSD, and heroin from our society would gain us far less than the elimination of tobacco use.

What should we eat or not eat to avoid cancer? The answer to that question is best given as three general principles.

(1) Eat a variety of foods. But variety need not include exotica. We are often told of secret foods known only to the ancient Greeks or Norsemen or some other admirable group and recently rediscovered by an obscure physician, scientist, or former celebrity and guaranteed to prevent or cure some disease or another. Utter nonsense: any modern supermarket provides an adequate variety of foods.

(2) As often as possible eat fresh, unprocessed foods. Foods preserved by salting or smoking should not be a regular part of the diet. On other hand, the presence of food additives does not necessarily make a food unhealthy; indeed, it may be improved.

(3) Shift your diet away from high-protein, high-fat foods and toward carbohydrates. Protein sources for non-vegetarians should be a reasonable balance between meat, fish, and skim milk. Carbohydrates should include as wide a variety as possible of fresh fruits, vegetables, and grains.

---

*A remarkable proportion of those who reject
today's treatments of cancer also seek our money.*

Is there a role for exercise in the prevention of cancer?
Perhaps so. Very recent epidemiological studies by John
Vena, Saxon Graham, and their colleagues suggest that cancer of the colon and breast occur less often in those with
lifelong habits of regular exercise. Hardly a sure thing at
this point, but their data do provide still another reason to
remain physically active throughout life.

Finally, a few words about those to whom you may lis-
ten politely, but whose advice you should never take. (1)
Those who scorn established approaches to cancer prevention and treatment. (2) Those who argue that "everything
causes cancer" and there's nothing to be done about it.

A remarkable proportion of those who reject today's
treatments of cancer also seek our money. A letter I received
in 1987 from G. Richard Hicks is illustrative. Mr. Hicks is
the Executive Vice President of the Linus Pauling Institute
of Science and Medicine.

Mr. Hicks begins with a fact: "This year 483,000 people
will die of cancer." Should that not induce sufficient fear, he
continues: "Many people will die in extreme pain after
lengthy doses of x-rays, radioactive substances, and a host
of powerful and debilitating chemicals. Cancer is the plague
of the 20th century." Then a question: "Are you concerned
enough about these alarming facts and the threat they represent to your family to support a research team that has made
startling progress toward the control of cancer?" Now the
message: "For a donation...you can support Dr. Linus
Pauling and his associates, who are making real progress
toward the control of cancer."

Should the name alone of "two-time Nobel Laureate Dr.
Pauling" not make you reach immediately for your check-
book, we are asked to "consider the recent accomplishments

(by him) and his principal associate in cancer research, the distinguished surgeon Dr. Ewan Cameron."

Mr. Hicks then tells us of the Pauling-Cameron finding that "terminally ill cancer patients receiving vitamin C lived an average of seven times longer than patients not receiving vitamin C therapy after reaching the terminal stage." Unfortunately, we've already seen in Chapter 14 that this "accomplishment" could not be replicated in two very careful attempts at the Mayo Clinic nor, judging by their failure to publish additional evidence, could it be replicated by Pauling and Cameron. It is essential that private citizens send money to Pauling because, in Hick's words, "The government's granting agencies are afraid of supporting pioneering research." Of course, Hicks provides no convincing explanation for this alleged fear.

The science of medicine can today provide no cure for most forms of cancer. It is for this reason that adequate funds for research are essential. Each of us, as taxpayers, support cancer research through the National Institutes of Health in general and the National Cancer Institute in particular. For those wishing to donate more, the American Cancer Society or your local hospice group are worthy recipients.

Now for the "everything causes cancer and there's nothing to be done about it" form of ignorance. An article written by the syndicated columnist, James J. Kilpatrick, is illustrative. Under the title "Is Everything Hazardous to Health?" Mr. Kilpatrick says that "...if cigarettes are hazards to our health, so is just about everything else under the sun."

Kilpatrick goes on to cite factors that somewhere, sometime, by someone have been associated with cancer. He lists bread mold, black pepper, hair dyes, hot soup, lack of iodine, synthetic fibers, lettuce, migratory birds, mushrooms, olives, peanut butter, tin cans, tomato juice, tooth paste, vodka, and wax cartons.

What Mr. Kilpatrick fails to note is the relative quantity and quality of evidence in support of the relationship

---

*Cancer is a family of diseases with no single cause.
We must then expect no single means of prevention.*

---

between cancer and any of the factors he lists. For cigarets, that evidence is massive and of high quality. Those who advocate tobacco smoke for themselves and for the rest of us while avoiding toothpaste, lettuce, and migratory birds appear to me to be either ignorant or in the employ of the Tobacco Institute. (Oddly enough it was the Tobacco Institute that compiled the list upon which Mr. Kilpatrick based his column.)

Summing Up

Organized medicine and the "cancer establishment" as exemplified by the American Cancer Society and the National Cancer Institute are often accused of emphasizing the technology of cancer treatment once established while ignoring simple means for the prevention of cancer. If that were true in the past, it hardly seems the case today. Indeed, a majority of the critical remarks in this chapter were directed at those who would too eagerly bring "tips for cancer prevention" to the American people in the absence of adequate scientific evidence.

Cancer is a family of diseases with no single cause. We must then expect no single means of prevention. However, on one point there can be no equivocation: chronic exposure to tobacco smoke causes cancer. In addition, because there are a number of industrial chemicals that may leak into our environment, we must insist that our government make strong and sustained efforts to protect us. As individuals, all of those aspects of diet and exercise that contribute to health in general may be expected to diminish as well our risk of cancer.

CHAPTER 22

Heart Attack and Stroke

The Basics

Because it is cholesterol that is deposited in our arteries [in atherosclerosis], cholesterol has come to be regarded as a villain. In fact, human life would be impossible without it. Cholesterol is a structural component of every cell of every animal.

Although cancer is more feared and AIDS has a more sinister face, it is plain old "heart disease" that kills the majority of Americans. More properly, it is atherosclerosis, a disease in which blood vessels become blocked by deposits of cholesterol (atheroma) and hardened (sclerosed). The vessels most often affected are the arteries, hence the related term arteriosclerosis, "hardening of the arteries." When it is the arteries that supply blood to the heart itself, a heart attack results; when it is the arteries of the brain, the consequence is a stroke.

Because it is cholesterol that is deposited in our arteries, cholesterol has come to be regarded as a villain. In fact, human life would be impossible without it. Cholesterol is a structural component of every cell of every animal. In addition, the adrenal glands and the ovaries use cholesterol as the first step in synthesizing essential hormones. Thus, the villain is not cholesterol; the villain is the disordered regulation of cholesterol that contributes to atherosclerosis. In this chapter we will consider some of the many factors that influence the development of atheroslerosis and its major consequences: stroke, heart attack, and death.

---

*If cholesterol is essential, how do strict vegetarians
survive? Without foods of animal origin, there is none
in the diet. The answer is simple: vegetarians and the
rest of us are perfectly capable of making our own.*

If cholesterol is essential, how do strict vegetarians survive? Without foods of animal origin, there is none in the diet. The answer is simple: vegetarians and the rest of us are perfectly capable of making our own. That's why there is nothing essential about dietary cholesterol. The primary site of synthesis of cholesterol is the liver. In addition, it is the liver that serves as the gatekeeper for all cholesterol entering the body via the diet.

Upon reading this last paragraph, we might guess that the origin of atherosclerosis is a faulty liver; it probably isn't. I say "probably isn't" because, despite medical research of Nobel Prize quality, the ultimate origins of atherosclerosis are not known and, despite the abundance of advice that is offered, we really don't know what to do about it. But there are a number of safe bets that can be made; a number of probably useful habits we can adopt. (In the last chapter I called them my superstitions: health-enhancing practices that are not fully proven, but in keeping with the best available knowledge.)

Oil and water don't mix. Cholesterol is an oily, water-insoluble material, and the blood is a watery medium. How then are we to move cholesterol from the liver to all the cells of the body? As any homemaker knows, to disperse oil in water, you need a detergent. Nature's cholesterol-dispersing detergents are the lipoproteins. They are of various kinds and are usually addressed by their initials; best known to the general public are LDL, low density lipoprotein, and HDL, high density lipoprotein.

LDL-cholesterol has come to be known as "bad cholesterol" because it the form most often associated with heart

> *...to disperse oil in water, you need a detergent. Nature's cholesterol-dispersing detergents are the lipoproteins.... best known to the general public are LDL, low density lipoprotein, and HDL, high density lipoprotein.*

disease. In contrast, HDL-cholesterol is not only not bad, it is believed by many to be a positive factor in the prevention of atherosclerosis. Both LDL and HDL transport cholesterol through the bloodstream. The difference is that LDL tends to move excess cholesterol into the places we don't want it, our blood vessels, whereas HDL ships it to the liver where it can be disposed of without harm.

The preceding sentence was easy to write, but in fact reflects a story that began more than 50 years ago and is still not ended. It is worth a few paragraphs. In 1938, Carl Muller, a physician at the Oslo Community Hospital in Norway, described a group of patients with an inborn propensity to develop atherosclerosis. The name given to their disease, familial hypercholesterolemia, reflects its genetic basis and its primary sign, elevated levels of cholesterol in the blood. About one person in a million receives defective genes from both parents and is said to be homozygous for the disease. These individuals are destined to suffer heart attacks while still in childhood. Despite its rarity, the study of familial hypercholesterolemia has provided insights likely to be of value to the majority of the world's population.

> *Both LDL and HDL transport cholesterol through the bloodstream....LDL tends to move excess cholesterol into the places we don't want it, our blood vessels, whereas HDL ships it to the liver where it can be disposed of without harm.*

> *Richard Mould [wrote] that the person least likely*
> *to develop atherosclerosis is a "hypotensive, bicycling,*
> *unemployed, hypo-beta-lipoproteinemic, hypolipemic,*
> *underweight, premenopausal, female dwarf living in*
> *a crowded room on the island of Crete."*

In 1972, Drs. Michael Brown and Joseph Goldstein of the Southwestern Medical School of the University of Texas set out to understand familial hypercholesterolemia. Others had earlier shown that it was cholesterol coupled to low density lipoprotein, LDL-cholesterol, that accounted for increased risk of heart attack so it was upon the regulation of LDL that Brown and Goldstein focused. They soon found that the cells of patients homozygous for familial hypercholesterolemia lacked receptors for LDL-cholesterol. As a result, cholesterol cannot be absorbed in a normal fashion, blood levels of LDL-cholesterol increase, there is deposition of excess cholesterol in the arteries, and arteriosclerosis results. These discoveries on the origin of atherosclerosis earned for Brown and Goldstein the 1985 Nobel Prize for Medicine.

What with HDL- and LDL-cholesterol, defective genes, and Nobel prizes, you may already have guessed that cholesterol and what to do about it are a little more complicated than one egg a day instead of two and eating oat bran muffins. If nothing else we need to figure out how to decrease LDL-cholesterol while increasing the HDL form. Fortunately, we now know how to influence both in the right direction.

If the primary problem in atherosclerosis is a genetically determined defect in LDL-receptor synthesis, then why all this talk about dietary cholesterol? The answer is simple: whatever our inborn ability to handle cholesterol in a healthy fashion, we can stress the system in various ways and thus contribute to atherosclerosis. If we happen to be that one person in a million who is homozygous for familial hypercholesterolemia, we have a tough road ahead: a very restricted diet and rigorous drug treatment with no guarantee of success. If we are genetically blessed, no amount of cholesterol in the diet will be hazardous to us. Most of us lie somewhere between. Most of us will benefit from a modest increase in HDL-cholesterol and a modest decrease in the LDL form, but very few have any need for draconian measures such as those advocated by the late Nathan Pritikin.

We will come back to cholesterol in a moment, but let's first consider the concept of risk factors. Broadly defined, a risk factor for heart disease is any feature of human life that is correlated, positively or negatively, with heart disease. We have already considered one risk factor: an elevated level of LDL-cholesterol. In 1981, Drs. Paul Hopkins and Roger Williams assembled a list of 247 such associations. Richard Mould was moved to write that the person least likely to develop atherosclerosis is a "hypotensive, bicycling, unemployed, hypo-beta-lipoproteinemic, hypolipemic, underweight, premenopausal, female dwarf living in a crowded room on the island of Crete." Her male counterpart is "an ectomorphic Bantu who works as a London bus conductor, spent the war in a Norwegian prison camp, never eats refined sugar, never drinks coffee...and is taking vast doses of estrogens to check the growth of his cancer of the prostate." Over the past decade or so, the media have solemnly told us of most of the 247 risk factors for atherosclerosis, but too rarely have put them in perspective. The reaction of many is to ignore the whole business; medical scientists are obviously a confused lot.

[Mould's] male counterpart is "an ectomorphic Bantu who works as a London bus conductor, spent the war in a Norwegian prison camp, never eats refined sugar, never drinks coffee...and is taking vast doses of estrogens to check the growth of his cancer of the prostate."

---

*...risk factors are not infallible predictors
of heart disease and stroke*

We could, if we had the time and the money, go off to the clinics of Mayo or Cleveland or elsewhere and have a fair fraction of the hundreds of potential risk factors evaluated. With the possible exception of the terminally hypochondriac, it wouldn't be worth it. The reason is that risk factors are not infallible predictors of heart disease and stroke; we've all heard of the obese, cigar-smoking man whose idea of exercise was to watch bowling on television who dies accidentally at the age of 89 while learning to ride a dirt bike. But exceptions to the rules are not the stuff of epidemiology or of public health. Let's come up with a list of risk factors shorter than 247 and then see what we can do about reducing our chances of a heart attack or stroke. First of all, we will assume that we are without symptoms of any disease; the person who has chest pain upon exertion should immediately seek medical evaluation. The rest of us may wish to consider the following nine factors: 1, family history; 2, male gender; 3, age; 4, high blood pressure; 5, diabetes; 6, raised serum cholesterol; 7, obesity; 8, smoking; 9, lack of exercise.

Evaluation of the nine items on this list will identify in advance about 60% of those who, if they don't change their ways, will suffer a stroke or heart attack in the next five years. Among the remaining 40% are a lot of apparently fit people who will die unexpectedly. Most have read of the middle-aged man who drops dead of a heart attack the day after being given a clean bill of health by his physician; that's the bad news on risk factors. The good news is that our list of nine major predictors includes six—family history, male gender, age, smoking, obesity, and lack of exercise—that can be evaluated at no cost. The next two, high blood pressure and diabetes, are easily detected, often as part of free screening programs. These days one can even get a cholesterol check for a nominal fee and sometimes for nothing. If more extensive cholesterol evaluation is needed, this will be more expensive. On the other hand, most of us will need only a few determinations of cholesterol over a period of decades.

Looking at our list of nine risk factors in another way, we see that though we have no control over our family history of heart disease, our sex, or our age, we can influence all the rest, either for better or for worse. In addition, there is considerable interaction between factors four through nine. For example, a regular program of aerobic exercise (#9) is an essential element in any weight-loss program (#7), has a beneficial effect on blood pressure (#4), decreases the amount of insulin we need (#5), reduces levels of low density lipoprotein cholesterol (#6), and can serve as a substitute for the addiction of smoking (#8). One other thing about the risk factors on our list: The order of their importance cannot be stated. Though it is true that you are either male or not male, all the rest come in varying degrees.

In the next chapter I will offer you very specific advice regarding each of the major risk factors for atherosclerosis. However, before doing that, I think we should examine in a little more detail how we got to where we are in our thinking about just one of them, serum cholesterol.

Although much has been learned in the past two decades, the level of controversy and confusion regarding cholesterol is as high today as ever. The questions to be answered appear simple: (1) Can the incidence of death from heart attack be reduced by lowering levels of serum cholesterol? (2) If a lowering of serum cholesterol is beneficial, how is this best achieved? More specifically, what role does dietary cholesterol play in determining the amount of cholesterol in the blood?

At least since Muller's description of familial hypercholesterolemia in 1939, there has been no doubt that very high levels of cholesterol can lead to death by heart attack at an early age. In addition, a number of epidemiological studies have suggested that there is a continuous relationship between cholesterol levels and risk, i.e., the higher the level, the greater the risk and the lower the level the better. More than 20 years ago, based on admittedly incomplete evidence, the American Heart Association made its first recommendations to reduce dietary cholesterol and saturated fat.

Although the hypothesis that high cholesterol levels are related to heart disease was attractive, most recognized that proof was lacking. In July 1970, the Task Force on Arteriosclerosis made plans for massive epidemiological investigations designed to settle the matter. The Task Force rejected the idea of a simple diet/heart disease study in healthy people on the grounds that the expected effect would be too small to be identified except in an unreasonably large group of people. Instead they opted for designs in which persons at high risk would be treated in various ways and the outcome of those treatments determined. Financial support for the investigations would come from the National Heart and Lung Institute (NHLI; later blood disease was added to its purview and it became NHLBI).

The Multiple Risk Factor Intervention Trial (MRFIT) was begun in 1972 and completed 10 years later at a cost of $115 million. The size of the bill reflects the fact that 361,662 men were screened for the presence of smoking, high blood pressure, and elevated blood cholesterol. Of those screened, 12,866, those at highest risk, were divided into two groups. The "usual care" group received an annual physical by MRFIT personnel and were referred to their own doctors. Those assigned to the "special intervention" group were checked every four months by MRFIT and efforts were made to get them to lower cholesterol by changes in diet, to stop smoking, and to lose weight. Elevated blood pressure was treated with hydrochlorothiazide and chlorthalidone, diuretic agents ("water pills") widely used in the early 1970s for the treatment of hypertension.

The MRFIT Research Group published their findings in September, 1982 in the *Journal of the American Medical Associ-*

ation. Although they tried to put a happy face on the report, the inescapable fact was that for every 1000 usual care patients 40.4 died and for every 1000 special intervention patients 41.2 died. Later analysis would suggest that special intervention had done some people some good. Unfortunately, the drugs chosen to treat high blood pressure seemed to kill more than they cured and, as a result, no net benefit could be shown.

NHLBI had more than one string to its bow. About a year after MRFIT was begun, a second study was organized to look more directly at cholesterol levels and the risk of coronary heart disease. It would be called the Lipid Research Clinics Coronary Primary Prevention Trial (LRCT). Nearly half a million men were screened for the trial to identify those whose total cholesterol put them in the top 5% in that regard (265 milligrams per deciliter or greater). From that group were selected those without other major risk factors and without signs of disease.

In other words, the participants were to be overtly healthy men whose only apparent problem was elevated cholesterol.

The 3806 LRCT subjects were assigned to either a treatment group, which received a cholesterol-lowering drug called cholestyramine, or a control group that was given a placebo. All subjects were placed on a diet expected to reduce cholesterol by 3–8%. The men were followed for an average of 7.4 years. The primary endpoint was death from coronary heart disease or a nonfatal myocardial infarction (heart attack).

The results of the Lipid Research Clinics study were published on January 20, 1984 in the *Journal of the American Medical Association*. Treatment with cholestyramine was found to have reduced the levels of cholesterol in the blood and that reduction was associated with fewer heart attacks and fewer deaths from heart attacks. What followed may be described as an explosion of media attention. *Time* magazine had a cover story called "Hold the Eggs and Butter" with the subtitle "Cholesterol is proved deadly, and our diets may never be the same." But as we've seen so many times before,

---

The simple message carried by the media was "the more you reduce cholesterol and fat in your diet, the more you reduce your risk of heart disease."...But the [study] wasn't a diet study, it was a drug study. And [it] didn't look at "average"...or even "normal" people...

---

the passage of information from the literature of science, with all its reservations and doubts and caveats, to the purity of the popular press is not always without distortion.

Let's look at the numbers: Some 32 cholestyramine-treated men died of heart of attacks versus 44 controls; 12 lives were saved, an impressive result. But there are other ways to die than by heart attack. Overall deaths in the cholestyramine group numbered 68, whereas 71 controls died. The difference is less impressive and, by the usual criteria of science, quite possibly the result only of chance. The thought crossed more than one mind that the drug treatment, though saving lives by reducing cholesterol, might be taking lives by some unknown toxic effect. To this day, no satisfactory explanation for the increase in nonheart disease deaths in the drug-treated has been provided; oddly enough, the excess deaths were caused by accident and violence.

The simple message carried by the media was "the more you reduce cholesterol and fat in your diet, the more you reduce your risk of heart disease." In fact, those were the words of Basil Rifkind, MD, of NHLBI and director of LRCT, as quoted by *Time* magazine on March 26, 1984. But the LRCT wasn't a diet study, it was a drug study. And the LRCT didn't look at "average" people or even "normal" people; it was concerned with white, middle-aged men with an average cholesterol level of 292 mg per deciliter of blood.

With a combined investment of a quarter billion dollars in the Multiple Risk Factor Intervention Trial and the Lipid Research Clinics Coronary Primary Prevention Trial, NHL-

> *...it is not only interesting, but also useful to know
> what lies behind what we think we know
> about nutrition and fitness.*

BI had to do something with the results. On December 10, 1984, it and the NIH Office of Medical Applications of Research convened a Consensus Development Conference to consider "lowering blood cholesterol to prevent heart disease." Drawing heavily upon the Lipid Research Clinics Trial, the panel made a number of recommendations. Most were of the flag, motherhood, and apple pie variety: If your cholesterol level is very high, reduce it and get rid of other risk factors such as smoking; if you are obese, lose weight; educate the public and their physicians about cholesterol. But the panel's point #3 had not been said before by a government agency: all Americans aged two to 90, should reduce dietary fat intake to 30% of calories, saturated fat to less than 10% of calories, and daily cholesterol intake to 250–300 mg or less.

This book is devoted to the proposition that it is not only interesting, but also useful to know what lies behind what we think we know about nutrition and fitness. After examining in some detail the Multiple Risk Factor Intervention Trial, the Coronary Primary Prevention Trial, and the NHLBI Consensus Conference on cholesterol and heart disease, you may be inclined to the view that the advice offered to the general public has far outrun the scientific evidence. Doubters are in good company. M. F. Oliver, Duke of Edinburgh Professor of Medicine at Scotland's University of Edinburgh, chose this title for his article in the *Lancet* for May 11, 1985: Consensus or Nonsensus Conferences on Coronary Heart Disease. Professor Edward Ahrens of Rockefeller University said that recommendations "should be based, not on faith or zeal or alarm, but on hard scientific evidence" and added that "existing evidence is far from convincing." Less restrained

---

*...what may be acceptable advice for sick people
may be bad advice for healthy people.*

---

than Drs. Oliver and Ahrens was Thomas Chalmers of the
Mt. Sinai Medical School: "I think (the NHLBI panel) made
an unconscionable exaggeration of all the data."

I am convinced that more than one detail of the present
campaign by NHLBI, The National Cholesterol Education
Program, will be proven to be bad advice. I am equally convinced that the overall plan of action is a good one. In the
next chapter we will examine the good and the bad and I
will end up endorsing a diet little different from that suggested by NHLBI. Before doing that, something needs be
said about what currently appears in the popular media about
heart disease, cholesterol, and what to do about them.

The information and advice that is offered is sometimes
bad, sometimes good, and sometimes unintelligible. I have
two major objections. First, if it ain't broke, don't fix it. By
that I mean that what may be acceptable advice for sick people may be bad advice for healthy people. I will use as an
example Robert Kowalski's book, *The 8-Week Cholesterol Cure.*

By age 41, Mr. Kowalski, a writer by profession, had a
serum cholesterol level of 284 mg/dL and had already had a
heart attack and two coronary bypass operations. His father
died at an early age from heart disease. It is clear that Mr.
Kowalski suffers from arteriosclerosis. Based upon his personal experience, his advice to all of us is to exercise (no spe-
cifics given), eat less fat and cholesterol, but more oat bran,
and take large doses of niacin, the B vitamin we talked about
earlier. Fat, cholesterol, and exercise are familiar themes, but
why oat bran and niacin?

In Chapter 7, I told you about the discovery in 1955 that
very large doses of niacin can lower cholesterol. Does this
mean that an elevated level of cholesterol in the blood is
caused by a deficiency of niacin? It does not. The effects of

---

*I am all for self-reliance in matters of health
but do-it-yourself cardiology makes about as much
sense to me as do-it-yourself brain surgery.*

---

niacin on blood cholesterol bear no relationship to its actions as a vitamin; nicotinamide, for example, has all the vitamin activity of niacin yet alters cholesterol not at all.

The subtitle to *The 8-Week Cholesterol Cure* is "How to lower your blood-cholesterol by up to 40 percent without drugs." Without drugs? When used in doses several hundred times the amount needed for vitamin activity, niacin is a drug with a full spectrum of adverse effects. The real virtue of niacin for do-it-yourselfers is that this drug can be bought in unlimited amounts in any vitamin store without the inconvenience of getting a prescription. I am all for self-reliance in matters of health, but do-it-yourself cardiology makes about as much sense to me as do-it-yourself brain surgery.

Niacin is not, as Mr. Kowalski suggests, a long-ignored cure for high cholesterol. Niacin is but one of several drugs that a physician experienced in the treatment of this disease might consider when other measures fail. Drs. Michael Brown and Joseph Goldstein, recipients of the Nobel prize for their work in this area of medicine, say this about niacin: "It should be used with caution and primarily in high-risk patients...who have not responded dramatically to dietary measures."

I don't think Mr. Kowalski did us a favor by promoting niacin as a cure for high cholesterol. Only time will tell, but it is quite possible that, as a result of reading his book, more will be harmed by the drug than helped. On the other hand, I am all for oat bran, in moderation.

In Chapter 3, I described how fiber came to be regarded as a desirable part of a healthy diet. For many years and in many minds, wheat bran has been equated to "fiber." But, as we saw in Chapter 3, fiber comes in many forms. Unlike wheat bran, oat bran is rich in soluble fiber, the kind that is fermented by the bacteria of the human gut. More than 30 years ago, Prof. Ancel Keys found soluble fiber in the form of pectin caused a modest decrease in cholesterol in the blood of mental patients at the Hastings State Hospital in Minnesota. Since then, evidence has steadily accumulated that one of the benefits of a high fiber diet may be lowered cholesterol.

The scientific basis for Mr. Kowalski's advice to eat oat bran is provided in large measure by the work of James W. Anderson, Professor of Medicine at the University of Kentucky. A 1984 study by Dr. Anderson gives us a good indication of what we can expect from oats. Twenty men with cholesterol readings ranging from 272 to 392 mg/dL experienced a 19% drop in those values after eating 100 grams per day of oat bran for 3 weeks. (Perspective may be gained by converting that much oat bran to Cheerios' equivalents; it comes to 50 one-ounce servings each and every day; when eaten with skim milk, they would provide about 7500 calories.)

Finally, it must be acknowledged that not all have been able to replicate Dr. Anderson's findings. For example, a report published in 1990 threw considerable cold water on the notion that oat bran has a specific cholesterol-lowering effect. Janis Swain, Frank Sacks, and their colleagues at Brigham and Women's Hospital in Boston found that an 87 gram per day supplement of high-fiber oat bran had no more effect on cholesterol than a low-fiber refined wheat product. They concluded that any effect of fiber on cholesterol probably arises from simple replacement of fat by fiber. In an accompanying editorial, William E. Connor, MD, put it this way: "...people who eat large quantities of oat bran or other, similar cereals for breakfast have little room for bacon and eggs..."

I do not wish to minimize the potentially beneficial effects of soluble fiber on serum cholesterol. I do wish to emphasize that oat bran is not the only source. For example, one of Dr. Anderson's studies found dried or canned beans worked as well as oat bran in reducing cholesterol. Other

---

...what appears in many of the popular media concerning atherosclerosis is the tendency to focus on a single factor: Change your diet or stop smoking or reduce your blood pressure or lose weight or meditate or start exercising or... Nonsense. Everything that we know about risk factors shows that they act in concert.

---

sources of soluble fiber include many fruits and vegetables; especially rich are bananas, oranges, citrus fruits, cabbage, carrots, grapes, potatoes, and apples; fiber and other good things tend to be concentrated in the skins; when they're edible, eat them. There is also brown rice, barley, and peas and beans (legumes) of all kinds to be selected from. Finally, I must also point out that we have only the foggiest notion of how this cholesterol reduction comes about and what the contribution of specific kinds of dietary fiber might be. As I said earlier: Variety, variety, variety, that's what is needed.

My second objection to what appears in many of the popular media concerning atherosclerosis is the tendency to focus on a single factor: Change your diet or stop smoking or reduce your blood pressure or lose weight or meditate or start exercising or... Nonsense. Everything that we know about risk factors shows that they act in concert. Consider just three: smoking, blood pressure, and serum cholesterol. Data from the Multiple Risk Factor Intervention Trial tell us that men aged 35–57 who don't smoke, have cholesterol less than 250 mg/dL, and diastolic blood pressure less than 90 will die from coronary heart disease in the next 5 years at a rate of 2.40 per 1000. High blood pressure increases the rate to 3.86; smoking to 5.62; elevated cholesterol to 6.12. But for the male who smokes and has high blood pressure and has too much cholesterol in his blood, the rate of death increases to 17.49; in just five years, he is more than seven times as likely to die of heart disease. The lessons are obvious:

(1) Each of us must consider the major risk factors in our own lives. Early in adult life, find out whether your blood pressure is elevated, whether you are diabetic or pre-diabetic, and whether your serum cholesterol is higher than it should be. Periodically throughout your life, have these factors checked again. (There continues to be no consensus in the medical community about how often overtly healthy people should be seen by a physician. In 1987, the American Heart Association suggested once every five years up to age 60, every two years between 60 and 75, and annually thereafter.)

(2) It it ain't broke, don't fix it. What is good advice and prudent medical treatment for the person who has just survived quadruple coronary artery bypass surgery may not be good and prudent for all of us. It may simply be irrelevant and unnecessary; and often it poses very real risks.

(3) Look at the whole picture. A number of modest changes in the patterns of your diet, your exercise program, and your lifestyle are easier to accomplish and more likely to be of benefit than a semiannual jumping from one "change-your-life," "medical-makeover," "live-forever" fad to another.

Summing Up

From the several hundred risk factors that have been associated with an increased incidence of heart attack and stroke, cholesterol has emerged triumphant. This is good, but also bad. The goodness arises from the correlation of elevated levels of cholesterol in the blood with increased risk: It clearly behooves all of us to learn of our cholesterol levels. The darker side of media fascination with cholesterol is that we may lose site of the fact that atherosclerosis is a multifactorial disease. The sedentary hypertensive obese diabetic chronic smoker who gives up eggs for breakfast is likely to be disappointed with the results.

CHAPTER 23

Heart Attack and Stroke

The Role of Diet and Exercise

High blood pressure, diabetes, elevated serum cholesterol, obesity, and smoking are certainly a disparate lot...yet each contributes to a single disease, atherosclerosis, and each may be significantly influenced by diet and by exercise.... a combination...more effective than either alone.

After seeing the title of this book and the chapters it included, an editor told me that it would never work; she said the topics were too disparate. My dictionary defines disparate as "completely distinct or different in kind; entirely dissimilar." The error in her thinking is nicely illustrated by the major risk factors for atherosclerosis. High blood pressure, diabetes, elevated serum cholesterol, obesity, and smoking are certainly a disparate lot with quite distinct origins and overall consequences, yet each contributes to a single disease, atherosclerosis, and each may be significantly influenced by diet and by exercise. Furthermore, a combination of exercise and diet, disparate as they may appear to be, is more effective than either alone. In this chapter we will consider the simple things that all of us can do to reduce our personal risks for suffering heart attack and stroke, using the nine factors outlined in the previous chapter.

FACTORS #1–3: FAMILY HISTORY; MALE GENDER; AGE. Although we can't change any of these, we certainly can make use of

...all of us as we age need to pay greater attention to the
possibility of atherosclerosis and its consequences....
family history provides the clearest picture of our future.

them. In a way, it's the converse of the principle of "if it ain't broke, don't fix it." If you are male, you do need to concern yourself more about heart disease than if you are a premenopausal woman. Likewise, all of us as we age need to pay greater attention to the possibility of atherosclerosis and its consequences. But it is family history that provides the clearest picture of our future. We've already considered the example of familial hypercholesterolemia in which those untreated can expect to die as teenagers and before. For most of us the indicators provided by our parents and grandparents, our aunts, uncles, brothers, and sisters will be much more subtle and easily modified. We cannot change our genetic endowment, but we can work to counter any disease to which endowment predisposes us. Tragically, some begin life with a favorable genetic profile and proceed to overwhelm it with a fully controllable factor such as smoking.

FACTORS #4–8: BLOOD PRESSURE; DIABETES; SERUM CHOLESTEROL; OBESITY; AND SMOKING. Each of these may properly be regarded as symptomatic of disease and each is subject to medical treatment. A sign of that medicalization is the use of powerful drugs in each of these disorders. As a pharmacologist, I suppose I should favor drug therapy. I don't—at least not until we've tried more natural approaches. The simple fact is that each of these contributors to atherosclerosis can be favorably influenced to a very significant degree in a majority of Americans without recourse to the two-edged swords of medical technology.

In the pages that follow we will examine some of the evidence in support of the notion that diet and exercise can not only prevent atherosclerosis, but also reduce the risk of premature death from heart attack and stroke. Although we will consider separately each of the risk factors, it will become clear as we go along that the same principles of exercise and of nutrition operate again and again.

Where shall we begin? In Chapter 20 we decided that it isn't essential to consult a physician before beginning a graded program of aerobic exercise, and it certainly isn't required when you make changes in your diet. However, in Chapter 20, I gave several reasons why it's a good idea to have a personal physician whom you trust and feel comfortable with. If nothing else, a visit to your doctor at the outset of your physical renovation will allow you to keep score better. Values for blood sugar, serum cholesterol, and blood pressure are useful indicators of what needs to be done and provide a baseline against which to measure our progress. Of course you already know whether you smoke and you have a fair idea of whether you are too fat; on the latter point, your doctor can add his or her professional opinion. Just as the program of aerobic exercise that I recommended in Chapter 19 is for everyone, the features of the diet that follows are, with the exceptions noted, likely to be of benefit to all. Intelligent eating and aerobic training need no justification beyond the fact that they make you feel good. If you already are in great shape physically, biochemically, and emotionally, you are probably either very young or are already doing the right things. But, for the purposes of this chapter, we will imagine a person who is borderline sick. His name is George and his baseline values are as follows: age, 40; blood pressure, 140/100; blood glucose (fasting), 130 mg/dL; serum cholesterol, 230 mg/dL; smoking status, one pack a day; weight, 20% over recommended and with an obvious roll of fat around the middle; occupation, new car salesman.

No single element of George's profile is terribly alarming. Perhaps one-third of his contemporaries have higher blood pressure. His blood glucose doesn't classify him as diabetic. Until very recently, many physicians regarded a serum cholesterol of 230 as requiring no action. A pack a day doesn't make him a heavy smoker and it's very uncertain what risks, if any, are imposed by being 20% overweight. But the whole is greater than the sum of its parts; George is a prime candidate for too early death from atherosclerosis. What action should George take? The temptation is to try to do everything at once: start a new diet, exercise, lose weight, stop smoking. Certainly these are desirable goals in the long run, but they are best reached one step at a time. Exercise comes first. By reaching the goal, set in Chapter 20, of at least three sessions of aerobic exercise per week, George will achieve more than improved fitness. He will also lay an unshakable foundation for effecting other desirable changes in his life. In addition to the physical changes that aerobic exercise works, it invariably elevates mood and provides a sense of well being that will carry George up the steps before him.

Hypertension or high blood pressure has been called the silent killer because, undetected and untreated, it can produce life-threatening damage in apparently healthy persons. The organs most often involved are the kidneys, the heart, and the blood vessels. Blood pressure is nearly always given in terms of millimeters of mercury. Thus, 120/70 means that the peak pressure in the system, the systolic pressure, is equal to a column of mercury 120 millimeters high. The second number, the diastolic pressure, is a measure of the minimum pressure exerted by the blood on the arterial walls.

We've already seen that atherosclerosis is a multifactorial disease, influenced by many different elements of our lives; one of those elements is blood pressure. Were that not a sufficiently confused situation, hypertension is itself a multifactorial disease influenced by both genetics and environment. Despite such complexities, hypertension is worth our special attention because it is arguably the most important of the controllable risk factors. For example, there is some evidence that many people can get away with elevated levels of cholesterol so long as they are not hypertensive as well. Most important, many will be able to bring their blood pressure into a healthy range without resort to the expense, inconvenience, and possible adverse effects of drugs.

No absolute values define high blood pressure. Most authorities now focus on the diastolic pressure and use criteria such as those shown below:

Normal: less than 85
High normal: 85–89
Mild hypertension: 90–104
Moderate hypertension: 105–114
Severe hypertension: 115 and above

Despite this emphasis on diastolic pressure, the systolic value is not without importance and you may be regarded as hypertensive even when your diastolic pressure is less than 90. The condition is called isolated systolic hypertension. At present there is consensus neither on the likely consequences of this disorder nor on what ought to be done about it.

What can George expect from exercise with respect to his high blood pressure? In the old days, the advice given to those who had suffered a heart attack or had hypertension was to take it easy so as not to strain the heart. Middle-aged men and women were often sentenced to a life of invalidism. Now it's not uncommon for people to recover from a heart attack and go on to run marathons. It comes as a surprise to many to learn that "the old days" were little more than 20 years ago. Remember it was not until 1967 that the first article on "jogging" appeared in the medical literature and, still more recently, John Boyer and Fred Kosch of San Diego State University provided the first evidence that aerobic training might be useful in hypertension.

Drs. Boyer and Kosch chose for their study 45 sedentary men aged 35–61 who differed only in their blood pressure. The 22 men they called normotensive had values of 140/90 or less. (Today, a diastolic pressure of 90 puts you in a "moderate risk" category; another sign of how rapidly our thinking can change.) The readings for the 23 men defined

---

...it seems clear that aerobic training, even without
weight loss, has beneficial effects on blood pressure...

---

as having diastolic hypertension were 159/105. The exercise program consisted of jogging, in the original sense of the word, i.e., alternate walking and running, for 30 minutes twice a week for 6 months. Initially, the target heart rate was 60% of maximum and at 3 months was increased to 70% of maximum. (By present standards this is a modest program. In Chapter 20, I suggested 30-minute sessions three times a week at an intensity of 70–80%.)

At the end of 6 months, the control group had an average drop of 6 mm in diastolic pressure. Still more impressive was what happened to the hypertensives. Diastolic pressure declined in each of the 23. The average blood pressure reading after exercise was 146/93. In the years since publication of these results, many have confirmed their findings and most agree that aerobic exercise itself that accounts for the drop in blood pressure. Some uncertainty is introduced because many people lose weight after they begin to exercise and weight loss alone tends to reduce blood pressure. Weight loss seems not to have been a major factor in the subjects studied by Drs. Boyer and Kosch. Although there was an average loss of two pounds among the hypertensives, a few actually gained weight. But whether body weight went up or down, all experienced a decrease in diastolic blood pressure.

Though it seems clear that aerobic training, even without weight loss, has beneficial effects on blood pressure, the situation is much less settled with respect to exercise and blood sugar. This is not novel. The role of exercise in the control of blood sugar has been controversial at least since 1797, when John Rollo, whom we met in Chapter 4, advised his diabetic patients not to exercise. Since then there have been successive waves of enthusiasm for and condemnation of vigorous exercise by diabetics. A major portion of any current disagreement can be swept away simply by distinguishing, as we did in Chapter 4, between type I and type II diabetes.

Type I, or insulin-dependent, diabetics require the regular administration of insulin. No program of aerobic exercise should be undertaken by insulin-dependent diabetics without first consulting a physician familiar with their condition. For type II diabetics, and that includes about 90% of Americans who are called diabetic, and for "pre-diabetics" like George, no such restrictions apply. Unfortunately, no guarantees can be provided either. Though nearly everyone agrees that aerobic training improves insulin sensitivity, it is much less certain that glucose tolerance is improved. Until these matters are better worked out, George would be well advised to lose at least some of his excess fat; the beneficial effects of weight loss on fasting blood glucose are entirely clear.

In contrast with the uncertainty that surrounds the ability of aerobic training to favorably influence diabetes and its consequences, no doubt exists concerning exercise's effect on cholesterol. Back when the running boom was just beginning, one physician was so bold as to say that marathon runners never have heart attacks. That this is not true is less a derogation of the virtues of aerobic exercise than a humbling reminder that atherosclerosis in its most virulent forms can be resistant to all of our efforts.

The present intense interest in exercise as prophylaxis for heart disease is a simple extension of studies begun in the late 1940s by J. N. Morris and his colleagues under the auspices of the Medical Research Council of England. But it was not exercise that initially interested Dr. Morris. "Our main objective," he wrote, "was to seek for relations between the kind of work men do and the incidence among them of coronary heart disease." In the best known part of the investigations, the drivers of London's double-decker buses were compared to their conductors. In a day's work, the drivers sat while the conductors were in near-constant motion from front to back and from deck to deck of the bus.

---

*...epidemiological studies [suggest] that exercise
has little or no effect in lowering blood cholesterol.*

In a pair of articles published in November of 1953, Morris reported that bus conductors suffered heart disease less often and in less severe form than did the drivers. To explain this observation, the investigators presented the provisional hypothesis that "men in physically active jobs have a lower incidence of coronary heart-disease in middle age than men in physically inactive jobs." Several years later Morris and Dr. Margaret Crawford speculated that "habitual physical activity is a general factor of cardiovascular health in middle-aged men" and that "regular physical exercise could be one of the 'ways of life' that promote health in middle age."

It's hardly a practical approach to prevention of heart disease simply to tell everyone to be a conductor rather than a driver. If, as Morris and Crawford suggested, physical activity on the job is beneficial, might leisure-time exercise do as well? Of more fundamental importance, what are the biochemical and cellular bases for the beneficial effects of physical activity?

Even as Morris and his colleagues were studying the effects of physical activity on heart disease, evidence was accumulating that many other factors might also be involved. Because it is cholesterol that forms the artery-plugging deposits of atherosclerosis, that substance was of prime interest. Sure enough, reports began to appear in 1957 showing a good correlation between serum cholesterol and the occurrence of angina and heart attacks. But to the disappointment of some, the epidemiological studies suggested that exercise has little or no effect in lowering blood cholesterol.

More than two decades would pass before a solution to the puzzle of how exercise might influence heart disease without apparent effect on cholesterol would be suggested. It was called the HDL Hypothesis. In the last chapter we

---

"How much exercise is enough?" The answer is not known with certainty...[but one study] found a threshold for beneficial effect on HDL to be 10 miles per week of running or its equivalent.

---

talked briefly about the detergent-like lipoproteins and their role in moving cholesterol through the blood. We spoke of high density lipoprotein (HDL) and low density lipoprotein (LDL). In the first issue of *Lancet* for 1975, brothers G. J. and N. E. Miller proposed that the development of atherosclerosis results primarily from low levels of HDL. They cited the Eskimos of Greenland who, despite high total and LDL cholesterol levels, rarely have heart attacks; the blood of Greenland Eskimos is rich in HDL. A decade earlier it had been shown that cross-country skiers in Sweden had remarkably high levels of HDL. Perhaps, the Drs. Miller suggested, it is through an elevation of HDL that physical activity works its protective effect against heart disease.

The HDL Hypothesis gained abundant support in the years that followed its presentation by the Millers. It is now widely accepted that the ratio of HDL to LDL cholesterol is a better indicator of risk than is total cholesterol. Likewise it is widely accepted that aerobic exercise can favorably alter that ratio. But we must again ask the question, "How much exercise is enough?" The answer is not known with certainty, but the evidence before us may be disappointing for those who have convinced themselves that a gentle stroll around the block confers all the benefits of strenuous aerobic exercise.

Dr. Peter Wood and his colleagues at the Stanford University School of Medicine found a threshold for beneficial effect on HDL to be 10 miles per week of running or its equivalent. That's not too bad; the minimal program of 90 minutes per week I recommended to you in Chapter 20 comes to 10 miles at a modest 9 minutes per mile pace. But I must add that there is a difference between a threshold for effect and an optimal effect. Studies by the American physician and epidemiologist, Ralph S. Paffenbarger, Jr., suggest an optimal effect of exercise on longevity to require the equivalent of 20–30 miles per week of gentle running. That vast majority of Americans, myself included, who are unwilling to double or triple the time devoted to aerobic exercise can take solace in the fact that much remains to be learned about the relationships between exercise, HDL, longevity, and a myriad of other factors. At least for now, we need feel no guilt in sticking with the minimal program recommended here.

We've seen that George can favorably influence his blood pressure, cholesterol level, and perhaps his blood sugar by a program of aerobic exercise. Indeed, some who begin with numbers similar to George's need do nothing other than take up regular aerobic exercise. On the other hand, simple changes in diet are so easy to make and so likely to be beneficial with respect to heart disease and stroke that George—and the rest of us—are unwise not to make them. We will begin with George's present eating habits, suggest specific changes that should be made, and then look at some of the evidence making us confident we're headed in the right direction.

Five days a week George begins the day with bacon, scrambled eggs, toast, two cups of coffee, and a doughnut. He likes to think of this as his country breakfast, fueling him up for the rigors of the day to come. Lunch is at Wendy's, Burger King, or McDonalds. George's wife is responsible for dinner and, knowing him to be a meat-and-potatoes man, invariably features those two items. The day ends with a bedtime snack, usually a luncheon meat sandwich with a glass of milk. George doesn't drink much, usually a few beers after work. Weekends are less structured, but often include a Saturday night steak at a favorite restaurant, Sunday brunch at his club, and, the highlight of the week, a late Sunday dinner at home with his family.

By the light of present knowledge, there is much wrong with the way George eats. Some of those who speak to the public about nutrition would scrap the whole program and have George eating only fruits, nuts, and vegetables; all preferably raw. Fortunately for George and the rest of us, such extremism is neither necessary nor justifiable; we needn't give up the pleasures of eating in order to eat well. (Much dietary nonsense seems based in what might be called Nutritional Calvinism, the belief that, if it tastes good, it must be bad.)

In recent years general dietary recommendations have been made by various medical and governmental groups. Whether the country of origin be England or the United States or Sweden or Canada or elsewhere, whether the advice be directed at the prevention of obesity or hypertension or heart disease, there is remarkable uniformity. It is only in detail and in emphasis that the suggestions differ. Thus, the points listed below represent an international consensus:

1. Eat less saturated fat and cholesterol. Because animal products are the primary sources of saturated fat and the sole sources of cholesterol, a simple way to do this is to become a vegetarian. Should that not appeal to you, several other things can be done. Eat less pork and beef; substitute fish and poultry (without the skin). Instead of butter, use margarine made from a polyunsaturated oil (sunflower and corn oils are the most common). Cook in vegetable oil rather than in butter or fat. Decrease the intake of cream, whole milk, cheese, and eggs; substitute skim milk and skim milk products. Avoid foods of unknown composition such as commercial baked goods, sausages, and other prepared meats.

2. Eat more carbohydrates in the form of fruits, vegetables, and grains. Starchy foods such as potatoes, rice, pasta, bread, and breakfast cereals are just fine.

Before we apply these suggestions to George's diet, two things need to be said. First, there are no absolute prohibitions; there are no "nevers"; eating a few of your mother-in-law's buttery holiday cookies does not mean instant athero-

---

*An army of nutritionists has been telling us for the
past 50 years that we shouldn't skip meals.
For children, the elderly, and the undernourished
in general, that is usually good advice. For the rest
of us...lunch is an eminently disposable item.*

---

sclerosis. Second, the extent to which we make difficult changes in our selection of foods should be influenced by need; this is still another way to say "If it ain't broke, don't fix it." For example, some among us are able to eat a dozen or more eggs a week, and maintain a level of blood cholesterol that is healthy by all standards. To exhort these individuals to cut down on cholesterol intake by giving up eggs is silly.

Looking back at the baseline values for George, it is obvious that he is not one of those rare individuals who seem able to flout every rule of good health. Despite relative youth, his blood pressure is elevated, the levels of sugar and cholesterol in his blood are too high, and he is overweight. Aerobic exercise will help but, having now looked at George's weekly menu, it is clear that some changes are needed there as well. In fact, George's diet is so bad that I feel obliged to suggest some rather drastic alterations.

Let's begin with George's "country breakfast" of bacon, eggs, toast, coffee, and a doughnut: too many calories, too much saturated fat, and no fruits or vegetables. The same items can of course be improved by making the bacon crisp, cooking the eggs without fat, and spreading the toast with a polyunsaturated margarine (nothing will much help the doughnut). But, at least for now, we will go all the way and substitute a bowl of a high-fiber cereal with skim milk and a glass of orange juice. The coffee can stay, but a nonsaturated-fat, nondairy creamer should be used. Aside from the possibility of headache upon withdrawal, dependence on caffeine appears to be innocuous.

An army of nutritionists has been telling us for the past 50 years that we shouldn't skip meals. For children, the elderly, and the undernourished in general, that is usually good advice. For the rest of us, especially if we eat like George, lunch is an eminently disposable item. Many have found that a walk or another form of mild exercise substitutes very nicely. For those who frequent an outlet of one of the fast food chains, the consequences are very much dependent on the choices that are made. Some of the items are truly horrendous: One Wendy's Triple Cheeseburger provides about half the total daily caloric needs of a 120 pound woman and nearly 60% of those calories are in the form of saturated fat. At least in the short run, George should avoid all such temptations and either skip lunch entirely or substitute low-fat, high-carbohydrate items from home.

George has complete control over the composition of his breakfast and lunch: He is free to accept my suggestions as he wishes. Dinner is another matter; he needs the cooperation of his wife. For that reason, no drastic changes will be made. Most dinners should begin with a small salad made almost entirely of fruits and vegetables; no eggs, cheese, bacon bits, or other fatty items. A reduced calorie dressing or vinegar and oil is all right, but a sour cream or cheese dressing will work against a number of our purposes. Every dinner may include potatoes, but their form is important. A baked potato with the skin on and a modest amount of corn oil margarine is a good choice. Not so good are mashed potatoes prepared with butter and topped with beef or pork gravy. In keeping with our general advice, dinner should more often include poultry and fish with less emphasis on red meats. Until the visible fat around George's middle decreases, he should usually avoid dessert and limit its quantity when eaten.

For many who wish to change the way they eat, the hours between dinner and bedtime are the most troublesome—a daytime will-of-iron softens like ice cream in the sun. We will take up this issue again in Chapter 25, but for now will simply suggest that George's bedtime snack include skim rather than whole or 2% milk, and a sandwich heavy on lettuce, tomato, and the like, and light on meat. One of the better choices would be a thin slice of whole turkey breast. Most processed meats should be avoided; many provide 60% or more of their calories as saturated fat.

Weekends can be a particularly difficult time. More food is eaten out and we have less control over its quality and quantity. The first rule is to maintain structure. By that I mean that George should eat his usual breakfast and lunch (or skip lunch) just as on a weekday. Even though Saturday night dinner is eaten out, it should incorporate the same principles as during the week; that is, it should emphasize fruits and vegetables and conscientiously minimize saturated fat. Unfortunately, it is impossible to obtain margarine in many restaurants because of laws favoring butter; you may thank the dairymen of your state for that legislation. Our second rule is to let one day a week be reward-day. For George, that comes on Sunday, a day to enjoy food to its fullest. Brunch and dinner with his family are as they have always been. For those of us who love to eat, there must be in every dietary scheme a day of reward.

Governmental programs designed to improve the lot of one group or another are as common as fallen leaves in the Autumn. Many such programs, like the leaves, simply dry up, blow away, and there is left only a slightly larger Federal deficit. In fact, they were doomed from the start; each contained a fatal flaw in design; no one had thought to ask how we would know whether they succeeded or failed. In matters of exercise and nutrition, many of us commit the same error of design. Over the years we collect a group of habits intended to keep us alive and well for a long time. A few have been proven to work: For example, don't smoke and you will be less likely to suffer lung cancer and cardiovascular disease. For most, we simply hope. Hope is a fine thing, but for George, whom we have asked to exercise and to make changes in his eating habits, we want something better. That something better is evaluation.

At the beginning of this chapter, I said that one of the reasons we wanted baseline values for blood pressure, serum cholesterol, blood sugar, and body weight was to let us keep score. In other words, to help us evaluate the effectiveness of the changes that we were about to make. Few of us are inclined, nor should we be, to continue activities without demonstrable benefit to us.

I can say with great confidence that George will be aware of the benefits of his program of aerobic exercise well before the initial three-month period is over. As the training effect comes on, so too will a sense of well being. With the training effect will also come increased confidence that he can take control of other elements of his life. It's for this reason that I have repeatedly urged establishing a sustained pattern of aerobic exercise before explicit attempts to lose weight or to stop smoking are undertaken. An objective assessment of the effects on George of the combination of regular exercise and the dietary changes I've suggested should take place after about a year. There is a good chance that the return visit to his doctor will reveal that his blood pressure, blood sugar, and serum cholesterol are all now in a healthy range.

In the land of diet books, everything works; the "I lost 79 pounds in just 12 days while eating all I wanted" kind of thing. The real world is a little different. For example, George may find that his laboratory values are good enough to satisfy his physician, but that, despite an obvious decrease in fat around the middle, his body weight hasn't changed at all; he really wants to lose five pounds. It is not time to run out and buy a diet book. Instead he need only make a minor adjustment in his weekly eating. For example, he might eliminate the bedtime snack, save 350 calories per day, and in about seven weeks the five pounds will be gone. Or imagine that his weight has dropped and everything else looks good, but his blood pressure is still in the high normal range. George's bedtime snack can stay, but his doctor may recommend paying more attention to his intake of salt. The principle in all of this is that we don't make changes just for the sake of change. Once regular exercise and a basically healthy diet become a part our our lives, changes are made with specific benefits in mind. Enjoy the things that do you no harm and save your energy for those that must be changed.

Now let's consider a little more science. Earlier I said that the advice to eat less saturated fat and cholesterol represents an international consensus. I exaggerated. The 17 countries that have issued dietary advice to their citizens are unanimous in their condemnation of saturated fat, but only one argues against cholesterol. We might guess that the other 16 have simply assumed that, since the animal sources of saturated fat and cholesterol tend to be the same, a reduction in saturated fat intake will automatically diminish cholesterol consumption. There's more to it than that. Despite the plausibility of the assumption that the amount of cholesterol in the diet controls the amount in our blood, a majority of the world's authorities on such matters reject that assumption.

In 1913, Nicolai Anitschkow fed large amounts of cholesterol to rabbits; the rabbits developed atherosclerosis. Thus was born the notion, which soon became medical dogma, that eating cholesterol contributes to heart and vascular disease. But this simple idea is complicated by an equally simple fact: Rabbits are not humans. Most of us, in contrast with rabbits, are able to maintain appropriate amounts of cholesterol in blood despite large fluctuations in the quantity of cholesterol in the diet. Our bodies do this by controlling the amount of cholesterol that is absorbed, manufactured, and excreted. The net result is that our cellular and hormonal needs for cholesterol are met and excess is avoided. It is for this reason that all countries but one place little emphasis on dietary cholesterol. The exception is the United States.

The Federal government, as well as quasi-official organizations such as the American Heart Association, currently recommend that all of us reduce our intake of cholesterol to less than 300 mg per day. If that advice were taken seriously,

Although those who argue that all of us should eat less cholesterol are clearly ascendant...a more rational approach is to have your serum cholesterol measured while eating your usual diet. Only in this way can you know whether something need be done about it.

egg consumption would drop to zero; a single extra-large egg yolk contains about 300 mg of cholesterol. My advice that George eliminate bacon and eggs from his everyday breakfast has less to do with fear of cholesterol than it does with excess calories and saturated fat, of which more in a moment.

Just what exactly is the relationship between the amount of cholesterol that I eat and the amount that ends up in my blood? The answer is that we don't exactly know. Forty years ago, Professor Ancel Keys and his colleagues at the University of Minnesota reported that dietary intake of cholesterol between 250 and 800 mg per day had no effect on blood cholesterol in the apparently healthy men that they studied. Since then, the fruit of countless chickens has been fed to thousands of men and women in hundreds of good, bad, and indifferent investigations. And the controversy goes on. A reasonable approximation of the truth is as follows. (1) A very low intake of cholesterol reduces serum cholesterol in most people. (2) A very high intake of cholesterol increases serum cholesterol in most people. (3) Significant differences exist between people in the way they respond to dietary cholesterol. (4) As odd as it may seem, the amount of cholesterol we eat is far less important than the amount and kinds of dietary fat in determining the level of cholesterol in our blood and our risk of atherosclerosis.

Although those who argue that all of us should eat less cholesterol are clearly ascendant at this time, a more rational approach is to have your serum cholesterol measured while eating your usual diet. Only in this way can you know whether something need be done about it. To its credit, the National Cholesterol Education Program, launched in 1987, placed its emphasis on exactly that approach.

Is any harm done by eliminating eggs from the American diet? Relatively trivial objections can be raised. Chicken ranchers will suffer. Many of us will be denied, in the name of good nutrition, a pleasurable part of eating as well as an excellent and inexpensive source of protein. I've already suggested that for the vast majority of us, the sacrificial offering of eggs is unnecessary. But there is a more subtle and more substantial hazard. In heeding the advice of the well-intended, many will assume that by not eating eggs and other sources of cholesterol they assure themselves a normal level of serum cholesterol and a low risk of atherosclerosis. Too often, Nature's way of telling us that this assumption is invalid comes in the form of sudden death. There is no substitute for the measurement of serum cholesterol early in adulthood and periodically thereafter.

We have earlier considered investigations in which hundreds of thousands have participated. Yet it was an observation in a single patient that truly can be said to have set the course we now sail with respect to diet and heart disease. It was in the Spring of 1952 that a letter appeared in the *Journal of Clinical Endocrinology* from Laurance W. Kinsell of the Alameda County Hospital in Oakland. It was titled "Dietary Modification of Serum Cholesterol and Phospholipid Levels" and reported a single patient in whom the substitution of vegetable oils for animal fat caused cholesterol levels to fall.

In the next decade, Dr. Kinsell's remarkable observation was confirmed in at least five laboratories, most notably those of E. H. Ahrens, Jr., at the Rockefeller Institute, D. M. Hegsted at Harvard University, and the aforementioned Ancel Keys. Drs. Ahrens, Hegsted, and Keys remain today leading figures in the diet/fat/cholesterol/heart disease controversies.

Most are agreed that the amount of saturated fat in the diet is the single most important environmental factor in

...in adults, no harm can be attributed to a diet in which as little as 10% of calories is in the form of saturated fat.

determining serum cholesterol. Likewise, most are agreed that polyunsaturated fats may counter the effects of saturated fats. But the evidence that would permit us to move beyond these points of consensus to practical advice is far from complete. Why then am I willing to accept without reservation that a reduction in saturated fat is a good idea? The answer lies not in the potential benefits that may be gained, although these may be considerable, but in a much simpler observation: In adults, no harm can be attributed to a diet in which as little as 10% of calories is in the form of saturated fat. With the typical American consuming in excess of 40% of calories as fat, with perhaps two-thirds of the total being saturated, most of us have considerable room in which to operate.

In Chapter 2, the thousands of different fats were classified in terms of their degree of unsaturation. Animal fats tend to be solid at room temperature and highly saturated. Vegetable fats, also called oils, are usually of the polyunsaturated kind. In hundreds of studies in which human subjects have been placed on artificial diets, it has been demonstrated over and over again that saturated fat tends to increase serum cholesterol and polyunsaturated fat tends to decrease it. We don't know why dietary fat is of greater importance than dietary cholesterol in controlling serum cholesterol, but we certainly can make use of that fact.

For those at very high risk of atherosclerosis because of an elevated level of serum cholesterol, very careful attention to diet under the direction of a physician or registered dietitian is essential. For the rest of us, including George, I don't think it a good idea to try to be terribly quantitative about the kinds of fat in our diets. We really don't know what the ideal proportion of calories from fat is, but at least down to about 20%, less seems to be better. And we certainly don't know what the ideal ratio of saturated to polyunsaturated fat is. Given this situation, what we need is some general information and a set of guidelines to be followed.

As stated earlier, the simplest way to reduce saturated fat in the diet is to eliminate all animal products. Strict vegetarians (vegans) do just that and tend to have low levels of serum cholesterol. Very few need or wish to go that far. An easy first step is to cut down on the eating of red meat. An average adult American gets nearly one-third of his or her total dietary fat from red meat, principally beef. If carbohydrates are substituted for that single item, total fat calories will drop from 40 to 28% and the ratio of saturated to unsaturated fat will simultaneously move toward a more desirable value. The reason why red meat is singled out is that (skinless) poultry and fish tend to contain less saturated fat. In addition, there is some evidence that the polyunsaturated fats of fish may reduce the risk of heart attack and stroke by a mechanism independent of cholesterol; more on that soon.

The second most important source of saturated fat in the American diet is milk and milk products. Though no harm will be done by the reduction or elimination of red meat, I don't recommend giving up milk. The answer of course is to drink milk without the fat: skim milk. In this way, we get protein, calcium, and other desirables without adding to our burden of saturated fat. If you love the super-premium ice creams, you should be aware that they are very high in saturated fat; that's where the smooth texture comes from.

With so many people worrying so much about the foods they eat, it's become common for many processed foods to be advertised as "cholesterol free" or "made with pure vegetable oils." That's fine so long as we remember that dietary cholesterol has a minor impact on serum cholesterol when compared with the effect of saturated fat; a cholesterol-free product high in animal fat is no bargain. Likewise, as was discussed in Chapter 2, pure vegetable oil does not rule out coconut oil or another saturated vegetable oil.

So far, we've considered fats as being either saturated or polyunsaturated. In fact there is something in between: monounsaturated fats. The best example is olive oil, a pure fat of which 72% is monounsaturated. Epidemiological studies have for a long time suggested that certain Mediterranean populations, especially the Greeks and Italians, have a lower incidence of atherosclerosis than would be expected from the fat content of their diets. The answer to this puzzle seems to lie in the substitution among these peoples of monounsaturated fat, usually in the form of olive oil, for meat and dairy fats. It's not certain at present whether monounsaturated fat has a direct effect on serum cholesterol; most believe that its beneficial effects arise simply from displacement of saturated fats from the diet. In either case, the use of olive oil and other oils high in monounsaturates, such as peanut and sesame seed oils, in place of animal fats seems a good idea.

I said earlier that arteriosclerosis can begin very early in life. As a result, there has been considerable interest in applying dietary principles seemingly appropriate for adults to children and even to infants. For example, in the mid-1970s, many pediatricians, having been enlisted in the war against atherosclerosis by organizations such as the American Heart Association, urged mothers to put their babies on skim milk and other low-fat, low-cholesterol foods. That this is a bad idea is suggested by the fact that human breast milk provides about 50% of its calories as fat and contains about ten times the amount of cholesterol as whole cow's milk. There is good reason to believe that in the first few years of human life, optimal development, especially of the brain and nervous system, requires relatively high levels of fat and cholesterol. For this reason, the current recommendation of national and international pediatric organizations is that full-fat milk is best until the age of several years and that skim milk is not appropriate until age 5 and beyond. On the other hand, a gradual increase in the proportion of calories from fruits and vegetables may lay the basis for good habits of eating throughout life.

Whether...dietary insults in early life are ever fully remedied is a matter of current controversy. There is no doubt that a diet beneficial to overweight adults may do harm when uncritically applied to infants and children.

A vivid illustration of what may happen when parents assume that their infants and children are, dietarily speaking, simply small adults was provided by Michael Pugliese, MD, and his colleagues of the North Shore University Hospital in Manhasset, Long Island. Writing in the August 1987 issue of *Pediatrics*, they described seven of their patients, 7–22 months of age, who were not growing normally. The cause turned out to be very simple: The children had a less than adequate caloric intake; they were being subjected to low-level starvation. These were not neglected children. Indeed, the parents thought they were helping their infants to avoid the black beasts of pop nutrition: obesity, atherosclerosis, "junk food dependence," and "unhealthy" eating habits. The authors sum it up well: "It appears that society's obsession with being slim and trim and its fear of heart disease have resulted in another disease of poor growth and delayed development in infancy." Whether such dietary insults in early life are ever fully remedied is a matter of current controversy. There is no doubt that a diet beneficial to overweight adults may do harm when uncritically applied to infants and children.

So far we have talked about diet as a negative factor, with the main recommendation being a reduction in saturated fat. But by now most Americans have heard the idea that some nutrients may actively reduce the risk of atherosclerosis; the health foods stores are full of such claims. Most suggestions of this type are too poorly documented to deserve mention. Two exceptions are alcohol and fish oils.

Regular heavy drinking is a significant contributor to high blood pressure, stroke, heart disease, and premature

---

"...nonexercisers can maintain levels of HDL-cholesterol similar to those who jog regularly by ingesting three beers a day." Some call [it] "aerobic drinking."...When an idea seems to be too good to be true, the chances are it is.

---

death; on that point there is no controversy. But, beginning in the 1950s, epidemiological evidence began to accumulate that a moderate intake of beer, wine, or whisky is associated with a reduced incidence of heart disease. Today, nearly all take this relationship to be proven, but controversy continues regarding the way in which it comes about.

In the last chapter it was noted that the liver is the major regulator of cholesterol levels in the body. Here are produced not only cholesterol, but also high-density and low-density lipoproteins (HDLs and LDLs), the detergents that permit cholesterol to be transported by the blood. That alcohol might act upon the liver is not a new idea. Kirrhos is a Greek word meaning tawny orange in color. Tawny orange is the appearance of the nearly destroyed liver of an alcoholic dying with the disease we call cirrhosis. Might it be that moderate amounts of alcohol stimulate the liver to handle cholesterol in a manner protective against heart disease?

Earlier we considered the HDL Hypothesis proposed in 1975 by G. J. and N. E. Miller to explain the protective effect of aerobic training on heart disease. The discovery that HDL synthesis by the liver is enhanced by moderate amounts of alcohol established a connection between runners and drinkers. So neat and attractive was the picture that many were inclined to accept a conclusion drawn in 1983 by Dr. G. Harley Hartung and colleagues at the Baylor College of Medicine: "...nonexercisers can maintain levels of HDL-cholesterol similar to those who jog regularly by ingesting three beers a day." Some called the phenomenon "aerobic drinking."

When an idea seems to be too good to be true, the chances are it is. Those who would substitute beer for aerobic exercise should be aware that more recent evidence suggests that, although both prolonged aerobic exercise and alcohol elevate blood levels of HDL, their effects are not the same. Instead, alcohol is associated with a subfraction designated HDL_3, whereas exercise elevates the level of the subfraction HDL_2. The present majority opinion, and it is little more than opinion, is that only HDL_2 confers protection against heart disease. Disputes about the mechanism of action do not of course alter the epidemiological observation that moderate drinking is protective; the apparent beneficial effect may simply involve factors other than HDL.

In describing George at the beginning of this chapter, I mentioned that he usually has a few beers after work. Calories aside, should he continue to do so? In an editorial in the *New England Journal of Medicine* for March 29, 1984, Charles S. Lieber, MD, of the Mt. Sinai School of Medicine gave this answer: "...in a moderate drinker who has demonstrated the capacity to maintain intake at an acceptable level, there is no compelling reason to change a life style and eliminate a pleasurable and possibly beneficial habit." I agree, but must note that others do not; many believe that the risks of uncontrolled drinking far outweigh any potential benefits.

We are not likely soon to see advertisements by the liquor industry urging us to "drink for your heart's sake." On the other hand, advertisements do suggest that we buy capsules filled with fish oil. Unlike the cod liver oil that was forced down the throats of thousands of children in the 1940s, the idea has nothing to do with vitamins A and D. Instead, the thought is that the polyunsaturated fats peculiar to some fish oils will help us to resist the development of heart disease. To find the origins of this peculiar idea, we must go back more than half a century.

In 1935, Hugh Sinclair was a medical student in England. He developed the then radical notion that the increased incidence of heart disease in developed countries was caused by faulty nutrition. He was aware of the discovery of linoleic

---

Eskimos are virtually free of heart disease despite a very high intake of fat....[Moreover] when cut, Eskimos bleed longer than others!

---

acid by George and Mildred Burr and upon completion of his medical studies, came to the United States to learn more about essential fatty acids. As with so many others, Sinclair's scientific career was interrupted by World War II; but he was far from idle. Stationed in northern Canada, he was struck by the much lower incidence of cholesterol deposits in the natives compared with young British and New Zealand pilots being trained in the area. He noted that Eskimos are virtually free of heart disease despite a very high intake of fat.

In the decade following the end of World War II, Sinclair wrote a number of articles and letters that attempted to draw the attention of the medical and scientific communities to essential fatty acids and their possible role in atherosclerosis, cancer, and other degenerative conditions. Not many listened and those who did tended to focus on vegetable oils; it was in 1952 that Kinsell demonstrated a fall in cholesterol when vegetable fat was substituted for fat from animal sources. Sinclair's description of an Eskimo diet high in animal fat and devoid of fruits and vegetables had little appeal. (The Algonquian Indians are thought to have named their northern neighbors, Esquimawes, "those who eat raw fish.") Frustrated in his efforts to establish a department of nutrition at Oxford, Sinclair could do little more than quote Leonardo: "Truth is the daughter of Time and not of Authority."

Time, nearly 30 years worth, and two lines of investigation, one epidemiological and the other experimental, would indeed bring us closer to the Truth. Beginning in 1971, Jorn Dyerberg and Hans Olaf Bang made several trips from their native Denmark to the Umanak district on the West Coast of Greenland. They provided the first detailed comparisons of the diets and blood chemistries of Danes and Greenland

Eskimos. The contrasts were remarkable. Despite a high-fat diet in both groups, the Eskimos had lower levels of triglycerides and LDL and higher levels of HDL. Still more interesting, Dyerberg and Bang confirmed and quantitated an old observation: When cut, Eskimos bleed longer than others!

The work of Dyerberg and Bang was solid and impressive; a touch of the dramatic was added by Hugh Sinclair. In March of 1976 Sinclair was, as one Nobel laureate expressed it, "an eccentric Oxford don," and 70 years of age. Having obtained a frozen seal, he proceded to eat it, fish, and nothing else for the next 100 days. His only supplement was vitamin E. Sinclair lost 26 of his original 211 pounds, not a bad thing, but in other respects this "heroic and foolhardy diet" did not entirely agree with him. He became lethargic and suffered nosebleeds and spontaneous bruising. He did not write a bestseller called the "Eskimo Quick Weight Loss Diet."

The experimental evidence needed to make sense of the observations of Sinclair and Bang and Dyerberg and others had in fact been accumulating from the time of Ulf von Euler's discovery in the 1930s of prostaglandins. Some may recall the story told of these substances in Chapter 2. Slow but steady parallel progress in understanding essential fatty acids and prostaglandins culminated in the mid-1960s with the discovery of the connection between dietary fatty acids and prostaglandin synthesis. This was quickly followed by the elucidation of the chemical diversity of prostaglandins and of the functional roles of thromboxane and prostacyclin. By the 1980s, a plausible explanation of fish oils, bleeding Eskimos, and heart disease could be offered.

Let's look again at the process of atherosclerosis as it is presently understood. There are two complementary hypotheses: the lipid hypothesis and the platelet hypothesis. Until recently, most attention has been paid to the former; the idea that elevated levels of cholesterol in the blood set the stage for deposition of cholesterol-rich plugs in our arteries. Reduce cholesterol levels by ingesting less saturated fat and more polyunsaturated fat as found in vegetable oils and the risk of atheroslerosis will be reduced.

The platelet hypothesis suggests that more than mere elevation of cholesterol in the blood is involved in atherosclerosis. We all expect from first-hand experience that a small cut will soon stop bleeding; the blood coagulates at the point of injury. A family of blood cells, platelets, clump together as a part of this process. When coagulation occurs at the site of injury, it is entirely a good thing. But what if that clot were to grow into the interior of the blood vessel? Flow through the vessel would stop and tissues supplied by it would be starved of oxygen and nutrients; when the blood vessel is a part of the supply system for the heart, the result is a heart attack; in the brain, the result is a stroke. As many are aware, it is now common medical practice to treat heart attack and stroke victims with anticoagulants, "blood thinners," drugs intended to reduce the tendency of blood to form clots.

In first mentioning the platelet and the lipid hypotheses, I said that they were complementary. Many now believe that blockage of blood vessels in the disease of atherosclerosis is accelerated by high levels of cholesterol (the lipid hypothesis) and by an elevated tendency for clots to form (the platelet hypothesis). The reduced risk of heart disease in Eskimos may now be explained in two ways. (1) Fats from fish and marine mammals, like vegetable oils, are rich in polyunsaturates and thus reduce cholesterol levels (the lipid hypothesis). (2) Something in the Eskimo diet diminishes clot formation (the platelet hypothesis). We now know that the "something" is a family of polyunsaturated fatty acids, designated omega-3, found in high concentration in cold water fish and mammals; Professor Sinclair's seal is a good example.

Unlike aspirin, omega-3 fatty acids do not produce their anticoagulant effects by inhibiting the synthesis of thromboxane. Instead, they promote the formation of prostacyclin, a prostaglandin with anticoagulant properties. In this respect the omega-3 fatty acids differ from those found in vegetable oils (designated omega-6). The latter have no significant effect on prostacyclin formation or on the process of coagulation. The best known of the omega-3 family is eicosapentanoic acid (EPA or IPA or timnodonic acid). Eicosapentanoic acid lends its initials to a popular fish oil supplement called MaxEPA; more about such products in a moment.

In looking back over the preceding paragraphs, I suspect that some readers may by now have gotten the idea that fish oils are the greatest thing since sliced bread. This thought would be reinforced if you have read elsewhere of tests of fish oils in such diverse conditions as skin disorders, arthritis, and the vascular complications of diabetes. I think it's a fine idea to substitute fish for meat several times a week. I think it a lousy idea for more or less normal people to buy fish oil supplements; I think it even a poorer idea for sick people to do so.

It has been estimated that sales of fish oil supplements soon will reach half a billion dollars per year. There are several reasons why the consumers of that oil may not become Eskimo-like in their resistance to atherosclerosis. First of all, it is just plain foolish to think that fish oil alone distinguishes Eskimos from Danes or from the rest of us. Just as a high-fiber diet does not make you a South African Bantu, fish oil does not convey a primitive Eskimo life-style with long hours of hard physical activity. Fish oil does not influence genetics; studies of Eskimos adapted to a European diet reveal the continued presence of significant differences in blood chemistry. And then there is the matter of quantity.

The Greenland Eskimo diet studied by Bang and Dyerberg and Sinclair provides about 6 grams of eicosapentanoic acid (EPA) in the form of a pound or so of raw or boiled whale or seal meat. The better known supplements, such as MaxEPA, ProtoChol, and ProMega are said to provide somewhere between 180 and 500 mg of EPA per capsule. Getting the Eskimo dose of EPA would thus require as many as 30 capsules a day. At 30 capsules a day, one belch would draw half the cats in the county. Expense is another consideration; depending on the brand chosen, 30 capsules cost $1.35–6.75 at my supermarket in 1990. But of course, fishy breath and a $1500 a year are little enough to pay for cardiovascular health. What we really want to know is whether EPA works.

A few lines back I suggested that Eskimos may be different from us in more ways than their EPA intake. Perhaps more important, there is reason to believe that the protective effects of their diet may involve more than just EPA. In 1985, Daan Kromhout and his colleagues at the Institute of Social Medicine at the University of Leiden, the Netherlands, reported the results their study of 1088 men in the small town of Zutphen. Chosen at random in 1960 from Zutphen's population of 25,000, each man was evaluated in terms of serum cholesterol, blood pressure, smoking habits, body build, physical activity, and occupation. In addition, a careful dietary history was obtained. After exclusion of those with signs of heart disease and those outside the age range of 40–59 years, 852 men were left for long-term evaluation. Over the course of the next 20 years, 390 of the men died and the cause of death was recorded. A detailed statistical analysis was then made to determine the relationship, if any, between fish consumption and death from heart disease.

The main conclusion of the Dutch investigators was easy enough to understand: "Mortality from coronary heart disease was more than 50 percent lower among those who consumed at least 30 grams of fish per day than among those who did not eat fish." Bring on the fish oil supplements said some; the Eskimo diet has been proven said others. But let's look more closely at the results. Thirty grams of fish is a little more than an ounce; the Eskimos eat about a pound of fish and marine mammals daily. Furthermore, the men of Zutphen don't eat just "fatty" fish like mackerel, herring, salmon, and albacore tuna. As a result, their heart-protective "dose" of 30 grams of fish provides only about 0.4 gram of EPA; that's less than 10% of the Eskimo ration. The sellers of

---

...it must be emphasized...that smoking accelerates the progress of atherosclerosis at all stages of the disease.

---

fish oil supplements recommend that you buy and swallow 10 or more of their capsules each and every day. Dr. Kromhout and his colleagues say this: "On the basis of the epidemiological and experimental data available, it seems justifiable to include a recommendation for one or two fish dishes a week in dietary guidelines for the prevention of coronary heat disease."

My advice to George? Eat fish several times a week. Indeed, substitute fish for red meat as often as you like. Avoid the use of saturated fats such as butter in its preparation. Stay away from fish oil supplements and especially avoid cod liver oil; more than a few have poisoned themselves with vitamins A and D by the indiscriminate use of that oil.

Now that George is exercising regularly and paying closer attention to what he eats, we can do something about his smoking. It is lung cancer that kills more men and women in this country than any other form of cancer—perhaps 90% of these deaths are caused by cigarets. But in this chapter on heart disease and stroke, it must be emphasized as well that smoking accelerates the progress of atherosclerosis at all stages of the disease.

Some may wonder, given the well-established hazards of this terrible form of dependence and drug abuse, why we did not start with smoking. Two factors contribute to this decision: (1) The incremental harm done to George by an additional six months or so of his pack-a-day smoking are minimal. (2) The likelihood that George will permanently stop smoking, and that is our goal, is much increased by first establishing regular exercise and sensible eating. To understand why this is so, we need consider nicotine dependence.

Nearly all who are destined to smoke for several decades or longer begin as teenagers. The initial attraction may include elements of rebellion, a desire to be like their friends, or the romantic image of smoking fostered by the movie and tobacco industries. The primary reasons for continuing to smoke into adulthood and old age include none of these. Smokers keep smoking because of the pleasurable effects that it brings and because of the distress they experience when they try to stop; in pharmacological terms, they are physically dependent upon the nicotine contained in their cigarets.

When used repeatedly, drugs such as nicotine, heroin, and alcohol cause changes to take place in the brain that become apparent only when the drug is no longer used. We refer to this phenomenon as physical dependence and to the effects that occur as the abstinence or the withdrawal syndrome. The syndromes following heroin or alcohol are dramatic and unmistakable. In contrast, physical dependence upon nicotine produces signs of withdrawal that are of such subtlety that for many years they went unrecognized; the most important of these are feelings of tension and anxiety. Thus the chronic smoker who craves a cigaret is driven both by the positive pleasures provided by nicotine and by the need to relieve the negative effects of the withdrawal syndrome.

There are nearly as many ways to give up smoking as there are ex-smokers; what works for one person may not work for another. An absolute requirement is a sincere desire to quit. Some are able to do it on their own, others benefit from group support. Nicotine in the form of a physician-prescribed chewing gum may help. Why substitute nicotine in gum for nicotine in cigarets? The idea is to allow the smoker to unlearn all the nonpharmacological pleasures of the habit without the simultaneous burden of nicotine withdrawal. Nicotine intake may then be reduced gradually until a drug-

---

...regular aerobic exercise will provide an alternative source of pleasure and a sense of well-being that will go far to assure...success in stopping smoking.

As with so many other seemingly simple questions
with common sense answers, what we "know"
about stress and disease ain't necessarily so.

---

free state is reached. In George's case, the prior institution of regular aerobic exercise will provide an alternative source of pleasure and a sense of well-being that will go far to assure his success in stopping smoking.

Finally, in our make-over of George, we must consider the fact that he earns his living by selling cars. Would he be better off, at least as far as the risk of a heart attack is concerned, if he were in a less "stressful" occupation, that of a janitor perhaps? Reasonably alert Americans know the answer to that question, or at least they think they do: If you value your heart, stress should be avoided. Hans Selye, the father of modern stress research, said that those who learn to cope with stress will live to be 100. (Selye died at the age of 75). As with so many other seemingly simple questions with common sense answers, what we "know" about stress and disease ain't necessarily so.

In the mid-1950s, Meyer Friedman and Ray Rosenman, physicians at the Mt. Zion Medical Center in San Francisco, began studies that would alter medical thinking about heart disease and, in the process, give rise to a whole industry devoted to stress reduction. Drs. Friedman and Rosenman studied three groups labeled A, B, and C. In group A were those men (all their subjects were men) characterized by "(1) an intense, sustained drive to achieve self-selected but usually poorly defined goals, (2) profound inclination and eagerness to compete, (3) persistent desire for recognition and advancement, (4) continuous involvement in multiple and diverse functions constantly subject to time restrictions (deadlines), (5) habitual propensity to accelerate the rate of execution of many physical and mental functions, and (6) extraordinary mental and physical alertness." In contrast, group B men were "characterized by relative absence of drive, ambition, sense of urgency, desire to compete, or involvement in deadlines." Group C, now long-forgotten, was composed of unemployed blind men similar in behavior to group B, but judged as well to suffer from chronic insecurity and anxiety.

In an article published in the *Journal of the American Medical Association* for March 21, 1959, Friedman and Rosenman reported that the men of group A were seven times as likely as members of either of the other two groups to have signs of coronary artery disease. This was indicated by a history of heart attack, by the presence of chest pain (angina), or by characteristic abnormalities in the electrocardiogram. They suggested that "behavior type A" is unique to modern Western societies and that type A individuals might benefit from either a change in their environment or from new patterns of response to an unchanged environment.

Let us step back a bit. In 1628, William Harvey, discoverer of the circulation of the blood, wrote that "every affection of the mind that is attended with either pain or pleasure, hope or fear, is the cause of an agitation whose influence extends to the heart." A hundred years ago, William Osler told his students that angina is more likely to occur in those who "work at maximum capacity, incessantly striving for success in commercial, professional, or political life." Thus, the idea that heart disease, especially as manifested in angina, is somehow correlated with behavior is hardly new. Two factors contributed to the popularization of the concept of "type A behavior" and to the creation of what Charles Krauthammer has called "the myth of mellow." The first was the recognition in the early 1960s that traditional risk factors for heart disease could predict only a fraction of those who would suffer a heart attack. The second was Meyer Friedman.

In the nearly four decades since he began his clinical observations, Dr. Friedman has been a tireless promoter of the idea that type A behavior is a primary, if not the primary, cause of coronary artery disease. His 1969 book on the sub-

---

*Hans Selye wrote that "vigorous exercise represents
a stress with long-term harmful effects similar to those
of infections, trauma, and nervous tension." How shall
we reconcile Selye and Ralph Paffenbarger [who] has
presented what some regard as convincing evidence that
longer life results from regular and vigorous exercise.*

---

ject, *Pathogenesis of Coronary Artery Disease*, addressed a scientific and medical audience. Two later books were aimed squarely at the general public. *Type A Behavior and Your Heart* by Drs. Friedman and Rosenman appeared in 1974. Ten years later, a revised and updated version of the book, now coauthored by Friedman and Diane Ulmer, was called *Treating Type A Behavior—and Your Heart*.

Despite the intuitive attractiveness of the notion that we will all live longer if we just eliminate stress from our lives, practical application presents some problems. What, for example, shall we do with exercise? In Chapters 19 and 20, I emphasized a gradual and gentle approach to the training effect, but there is no denying the fact that exercise is an intentional stressing of the body. Hans Selye wrote that "vigorous exercise represents a stress with long-term harmful effects similar to those of infections, trauma, and nervous tension." How shall we reconcile Selye and Ralph Paffenbarger? The latter has presented what some regard as convincing evidence that a longer life results from regular and vigorous exercise. It is obvious that we need a clearer definition and understanding of "stress" before concluding that it is bad, everywhere and always.

But, exercise aside, what can be made of type A behavior? Following their initial definition of the term, Friedman, Rosenman, and their colleagues began what was called the Western Collaborative Group Study. After being classified as either type A or type B, 3154 men in California were followed for an average of about 9 years. The presence of type A behavior was associated with signs of coronary heart disease twice as often as was type B. In the eyes of many, the Friedman hypothesis had been confirmed. Others were not so sure. For example, the Multiple Risk Factor Intervention Trial that we discussed in the last chapter failed to detect any relationship between behavior patterns and heart disease.

Despite the widespread acceptance, by laymen as well as by physicians, that type A behavior is a health hazard, the foundations of that idea have continued to shake. For example, David R. Ragland and Richard J. Brand of the School of Public Health of the University of California at Berkeley wondered about the fate of the men identified in the Western Collaborative Group Study as having heart disease. Common sense might suggest an early death for the type A's.

During the initial period of observation in the Western Collaborative study, 257 men either had a heart attack or suffered the pain of angina pectoris. Drs. Ragland and Brand then extended the study to determine the 22-year mortality of that group of 257. Their results, published in the *New England Journal of Medicine* for January 14, 1988, surprised many. For those who died within 24 hours of the primary event, there was no difference between type A and type B men. For those who survived the first day, men exhibiting the type A behavior pattern were 50% less likely to die as were men of type B. Rather than leading to an earlier death, type A men lived longer. Ragland and Brand: "...the clinical implication is that an intervention to change type A behavior is not justified for purposes of secondary prevention after a coronary heart disease event."

Controversy will continue to surround the question of the influence of behavior patterns on heart disease for the foreseeable future. Many of those who have been converted to the Friedman faith will continue to believe in a simple relationship between type A behavior and heart disease. Others will point out that the type A pattern is not present in many who develop coronary vascular disease and labels many others as being at high risk who continue for years without developing the disease. My personal view of the matter is well represented by Stephanie Booth-Kewley. After a careful review of all of the evidence, she wrote the following.

"Overall, the picture of the coronary-prone personality... does not appear to be that of the workaholic, hurried, impatient individual, which is probably the image most frequently associated with coronary proneness. Rather, the true picture seems to be one of a person with one or more negative emotions; perhaps someone who is depressed, aggressively competitive, easily frustrated, anxious, angry, or some combination...a maladapted personality."

What then of George and his chosen career as car salesman? Should he strive to be type B, hoping to develop a "relative absence of drive, ambition, sense of urgency, desire to compete, or involvement in deadlines?" I think not. Instead he might wish to consider Colonel Saito's simple advice to Colonel Nicholson at the River Kwai: Be happy in your work.

Summing Up

Words such as "natural," "holistic," and "alternative" are commonly used these days. If nothing else, their popularity reflects a vague but widespread dissatisfaction with and mistrust of establishment medicine. In fact, diet and exercise and their influences upon us are natural and holistic and alternative in the best sense of these words. The suggestions that I have made to George in this chapter with respect to diet and exercise will reduce his chances of suffering a heart attack or stroke; they probably will allow him to live longer; these are not their only benefits. By beginning a life-long affair with aerobic exercise and thoughtful nutrition, George will enjoy more fully each day given to him. He will begin to appreciate the fact that health is more than a disease without symptoms; health is a positive state that all of us, so long as we are alive, enjoy in varying degrees; health is a condition that each of us, by our own efforts, can influence.

CHAPTER 24

The Prevention of Osteoporosis

---

*...approximately 200,000 hip fractures...each year
are suffered by elderly people...
The cost of all hip fractures exceeds $7 billion annually...
But consider the disease in human terms.
One out of five hip fracture victims is dead within one year.
Only slightly more fortunate [are the] hip-broken elderly who
will linger for years without ever again leaving their beds.*

In late 1986 the National Institutes of Health sought applicants to establish "Programs of Excellence of Research on Osteoporosis." In the words of NIH, "osteoporosis is a frequently occurring disease (or possibly several related diseases) that leads to weakened bone that fractures more easily than normal bone. The condition affects as many as 15–20 million individuals in the United States."

An appreciation of the consequences of osteoporosis has been slow in coming. NIH: "About 1.3 million fractures attributable to osteoporosis occur annually in people age 45 or older. Among those who live to be age 90, 32 percent of women and 17 percent of men will suffer a hip fracture, most due to osteoporosis combined with a fall. The vast majority of the approximately 200,000 hip fractures occurring each year are suffered by elderly people. The one-year mortality of hip fracture victims probably exceeds 20 percent. The cost of all hip fractures exceeds $7 billion annually."

In these days of $20 million contracts for quarterbacks, $50 million dollar fighter planes, and $3 billion space shut-

---

*...in severe osteoporosis, the bones may be so thin and
weak that fractures occur with the slightest stress;
a sneeze will sometimes suffice.*

---

tles, the dollar cost of osteoporotic hip fractures may not
impress you. But consider the disease in human terms. One
out of five hip fracture victims is dead within one year. Only
slightly more fortunate is that much larger percentage of the
hip-broken elderly who will linger for years without ever
again leaving their beds. (A "mattress grave" is the expres-
sion Heinrich Heine used.) To this we must add the countless broken arms and the collapsed vertebrae of "dowager's
hump" to begin to grasp the consequences of osteoporosis.

In Chapter 15, we considered the diseases called rickets
and osteomalacia (adult rickets). We saw that these conditions are characterized by soft bones because of inadequate
calcification. Rickets and osteomalacia are almost always
caused by vitamin D deficiency and the resultant inadequa-
cy of absorption of calcium. The disease called osteoporosis
is quite different. Bone appears to be normally calcified, but
there is simply too little of it; in severe osteoporosis, the bones
may be so thin and weak that fractures occur with the slightest stress; a sneeze will sometimes suffice.

Prior to World War II, those few who thought about
osteoporosis were inclined to attribute it to a lack of dietary
calcium. That is certainly the simplest explanation, but a half
century later its validity is still debated. Others of the 1930s
spoke of disuse, gastric hypoacidity, repeated pregnancies,
long-standing thyrotoxicosis, senescence, and Cushing's disease. In the May 31, 1941 issue of the *Journal of the American
Medical Association*, a fresh idea made its appearance. The
article by Drs. Fuller Albright, Patricia Smith, and Anna Rich-
ardson of the Massachusetts General Hospital was simply
titled "Postmenopausal Osteoporosis, Its Clinical Features."

Albright and his colleagues did not dismiss out of hand a possible role for calcium deficiency in osteoporosis, but they "were inclined to minimize the effects of diet." Instead they focused on the unquestioned observations that osteoporosis is more common in women than in men, and that it occurs in women only after the menopause or removal of the ovaries. Might then a relative or absolute lack of estrogen be the major causative factor in osteoporosis or at least in that branch of the disease they would call "postmenopausal?"

During the decade following World War II, the Albright hypothesis became thoroughly imbedded in the thinking of American physicians. Osteoporosis came to be regarded as a rather simple disease caused by lack of estrogen; a disease simply treated by providing estrogen to postmenopausal women. Many of you are already aware of course that so-called "estrogen replacement therapy" (ERT) is still used to treat and (one hopes) to prevent osteoporosis, but we are much less certain today regarding its benefits. Indeed we are much less certain today of the value of any preventive measure.

B. E. Christopher Nordin, a physician at the Gardiner Institute of Medicine in Glasgow, Scotland, in the late 1950s raised questions about the notion that osteoporosis is, pure and simple, a disease of estrogen deficiency. In effect he returned medical thinking to the pre-Albright era when osteoporosis was considered by some to be a disease of calcium deficiency. Dr. Nordin cited evidence of a lower intake of calcium by women who developed osteoporosis than those who remained free of the disease. He pointed out that estrogen replacement therapy had not yet been proven to be either curative or preventative. Nordin's own studies indicated that calcium supplements worked about as well as estrogen, though neither treatment could be shown to increase bone mass. For Dr. Nordin, the therapeutic implication was obvious: "patients should be treated with a high calcium diet."

Today, estrogen and calcium remain the foci of controversy. Simply put, estrogen replacement in post-menopausal

> *...estrogen replacement in post-menopausal women is*
> *probably of benefit to those 25% or so who need it,*
> *but we don't know in advance who needs it and the*
> *treatment has potential risks for all. On the other hand,*
> *calcium supplements of modest magnitude are unlikely*
> *to be harmful, but alas are of uncertain value to any.*

women is probably of benefit to those 25% or so who need it, but we don't know in advance who needs it and the treatment has potential risks for all. On the other hand, calcium supplements of modest magnitude are unlikely to be harmful, but alas are of uncertain value to any.

Most women find it much more convenient to regard osteoporosis as a disease of calcium deficiency than as something requiring estrogen replacement. The latter course means paying a physician to write an expensive prescription. And then there are the suspicions that estrogen may contribute to cancer. Calcium supplements require no doctor's prescription and are relatively inexpensive. Advertising makes the choice even easier. Estrogens are prescription drugs and by law cannot be sold directly. Calcium supplements are hawked in every magazine and newspaper, on every television set if not on every street corner.

A typical advertisement shows a bent old woman and straight young one. We are told that "eight out of ten women don't get enough calcium in their diets and, with time, that could lead to osteoporosis and its distressing consequences— weak, brittle bones and stooped posture." But have no fear, the ad goes on: "Osteoporosis, the brittle bones disease that plagues more than 20 million women, is in fact, a highly preventable condition." Osteoporosis, we are assured, can be prevented by calcium supplements. I can't dispute the effectiveness of ads like this; in 1989, nearly $200 million worth of calcium pills were sold. But a line by H. L. Mencken comes

*...in 1989...$200 million worth of calcium pills were sold....
a line by H. L. Mencken comes irrresistibly to mind:
"For every human problem there is a solution
that is simple, neat, and wrong."*

irrresistibly to mind: "For every human problem there is a solution that is simple, neat, and wrong."

The course of research into osteoporosis from Albright and estrogen in the 1940s through Nordin and calcium intake in the 50s and 60s to today is easy to follow. The reason is that, relative to other areas of disease, little has been done; there has been no "war on osteoporosis." Instead we have a few landmark investigations that are cited over and over in support of one position or another. Let's consider the most influential of these and then see if we can arrive at a rational program of prevention.

In 1965, a group led by Dr. Marc Moldawer at the Baylor University College of Medicine reported the results of their study of 100 black men and women over the age of 65. X-ray examination of the patients' spines revealed signs of osteoporosis, but with a frequency only about half that of comparable Caucasian men and women. When they examined the hospital records, it was discovered that hip fractures were five times as likely in white men and women as in blacks.

Moldawer's results were not entirely unexpected. Recall that A. R. P. Walker had used the lower incidence of osteoporosis in the Bantu of South Africa as a basis for rejecting the idea that dietary calcium is an important factor in the disease. Others had presented epidemiological evidence that blacks are somehow protected against the consequences of osteoporosis. The unique contribution of Moldawer was to suggest that bone mass and density might be the crucial factors. Women have smaller and less dense skeletons than men. Blacks have larger, more dense skeletons than whites. White women have the highest incidence of osteoporosis; black men

---

Women have smaller and less dense skeletons than men.
Blacks have larger, more dense skeletons than whites.
White women have the highest incidence of osteoporosis;
black men the lowest.
Prevention would then become a matter of figuring out
how to give white women the skeletons of black men.

---

the lowest. Prevention would then become a matter of figuring out how to give white women the skeletons of black men.

The idea that skeletal mass is an important factor in osteoporosis received support from Stanley Garn and his colleagues at the Fels Research Institute. In 1967, after studying 13,000 subjects in seven countries, some for as long as 23 years, they reached the following conclusions. Bone loss with age is a general phenomenon in both sexes, but women lose bone about twice as fast as men. Bone loss in women begins before menopause, but the process is hastened by menopause. Dietary calcium is unrelated to bone loss and intakes as high as 1500 mg/day do not protect against it. Finally, "all people ultimately lose bone, as far as we know, but those who have achieved a larger bone mass by the 4th decade are slower to evidence the clinical concomitants of bone loss in later life."

Upon considering the results of Moldawer and of Garn, some in the medical community thought the case of osteoporosis closed. A few went so far as to suggest that it isn't a disease at all, but instead a normal consequence of the aging process. Estrogen replacement therapy and increased dietary calcium as preventive measures decreased in popularity as the evidence became stronger that neither could reverse age-related (senile) or post-menopausal osteoporosis. These rather negative attitudes prevailed for a decade and indeed are still reflected today in some textbooks of medicine.

The origins of today's renewed enthusiasm for calcium supplements can be traced to two sources. The first was a study published in 1979 by Dr. Velimir Matkovic and his colleagues of the University of Zagreb. Although Zagreb is in Yugoslavia, the effort was international in character. One of the authors was the Englishman B. E. C. Nordin, who 20 years earlier had revived interest in dietary calcium as a preventive measure in osteoporosis, and the study was funded in part by the United States Department of Agriculture.

Yugoslavia is a country in which milk and dairy product consumption is quite high in some districts and quite low in others. This fact provided the investigators an opportunity to compare racially identical people living under similar environmental conditions who differed in their intake of calcium. Two districts in which calcium consumption was expected to be quite different were chosen and dietary surveys were done to confirm that expectation. In addition, the bone mass and density of the second finger of the left hand of residents of various ages were measured. Finally, a survey of hospital records was used to estimate the rate of bone fractures by age in the two districts.

As epidemiological investigations go, the results were remarkably clear-cut. In agreement with Garn, bone mass and density decreased with age in both districts and the rate of loss was not influenced by calcium intake. However, in the high-calcium district, they were higher at all ages. Most important, hospital records over a 6-year period showed a decreased incidence of hip fracture in the district with high calcium intake. These data return us to the idea that if we could increase skeletal mass early in life, age-related loss might not result in fractures. As I put it earlier: Give a white woman the skeleton of a black man.

The Yugoslav data certainly indicate that increased dietary calcium throughout life may protect against osteoporosis. But one study does not a conclusion make and other investigations equally well done have failed to show an influence of calcium intake on bone health. In addition, critics would later point out that the Yugoslav district with the higher calcium intake also had a higher energy intake, which in turn suggests a higher level of physical activity. For the second factor behind the calcium boom of the 1980s we must look outside of science to what I will call "nutrition activism."

In various parts of this book we encounter controversial issues in nutrition: sugar as a poison; diet and cancer; the role of dietary cholesterol in heart disease—the list is long. For each issue, evidence can be assembled on both sides. The traditional approach by responsible scientists has been to delay "going public" until the evidence overwhelmingly favored one position or another. In contrast, the nutrition activist believes that the public should be informed at a much earlier stage of discovery. In principle, one can hardly argue with that position. In practice, it often leads to misinformation and misunderstanding that is seized upon by a spectrum of quacks and charlatans eager to make a profit.

Perhaps the most influential calcium activist in the country is Robert P. Heaney, MD, Professor of Medicine at Creighton University in Omaha, Nebraska, whose studies and opinions have indeed been seized upon by many commercial opportunists. In November of 1982, Dr. Heaney and his colleagues published in the *American Journal of Clinical Nutrition* an article entitled "Calcium Nutrition and Bone Health in the Elderly." They reviewed in an even-handed fashion much of the evidence that we have just discussed and were properly cautious in most respects. However, their summary concluded that "available evidence is compatible with allowances (of calcium) of at least 1200 to 1500 mg per day" and "that daily intakes up to at least 2500 mg...are quite safe" for most people. It was their opinion that "calcium nutrition is considerably more important in the genesis of osteoporosis than has been commonly thought for the past 35 years."

You don't find the *American Journal of Clinical Nutrition* next to the *National Enquirer* at the supermarket checkout. We would expect the Heaney article to be read by few other than professional nutritionists. We might also expect that few converts to calciumism would be made; no new evidence had been presented and most readers of the *Journal* had been hearing the same arguments for at least 20 years. It would take something else to put calcium supplements in the mouths of millions of American women.

On April 4, 1984, the National Institutes of Health convened a "Consensus Development Conference on Osteoporosis." On August 10, 1984, NIH's Office of Medical Applications of Research (OMAR) published in the *Journal of the American Medical Association* the conclusions of that conference. Now it should be noted that OMAR is in the business of health activism; its mandate is to quicken the transfer of findings in medical research to the American public. Therefore, we should not be surprised that the language in its report is positive, upbeat, and ready-made for quotation by the news media. "After one and one-half days of presentations by experts in the field, a consensus panel...considered the evidence and agreed on answers." The cornerstone of the panel's answer to the question "How can osteoporosis be prevented?" was "an elemental calcium intake of 1000 to 1500 mg per day."

The NIH Consensus Panel on Osteoporosis of 1984 is a fine example of nutrition activism. A half century of scientific controversy is reduced to one and one-half days of "presentations by experts" and "answers" are agreed upon when in fact no answers are at hand. Complex issues and unproven hypotheses are hammered into simple recommendations for action. Not all were displeased with the results of the conference. The calcium activists like Dr. Heaney had been heard. And for the sellers of calcium supplements, the panel was a dream come true. To have the apparent backing of the NIH was a guarantee of success for the calcium vendors.

Let's now return to reality. Consensus is not truth; the medical community has "agreed on answers" to health problems before, only later to be proven wrong. In the case of osteoporosis, it wasn't even a consensus of the medical community. The panel said that the RDA for calcium is too low. The Food and Nutrition Board of the National Academy of Sciences does not agree; their only concession in the 1989 RDAs was to extend the recommendation for 1200 mg per day from ages 11–18 to ages 11–24. The consensus panel thought it reasonable to publish what is essentially a broad recommendation for the use of calcium supplements. Others would suggest that such a recommendation would surely lull the public into believing that a cure for osteoporosis was nearly at hand.

In fairness to the panel, it must be noted that in addition to calcium supplements, they mentioned estrogen replacement in postmenopausal women and a program of "modest weight-bearing exercise." Unfortunately they provided no details about what exercise might be useful or even what is meant by "weight-bearing." As for estrogens, the reluctance on the part of many physicians to prescribe these drugs and of women to use them was reinforced by the panel's implication that a high calcium intake could substitute for estrogen replacement. (General Mills publishes a monthly newsletter called *Contemporary Nutrition* that it sends free to dietitians and other professionals. In the May 1983 issue, Robert R. Recker, MD, characterizes dietary calcium as "a safer alternative" to estrogen replacement.)

The starting point of any rational program for the prevention of osteoporosis is recognition that we don't know how to prevent osteoporosis. In this situation the best that we can do is to try to cover all the bases without trading doubtful benefit for probable harm. To do that, we must consider more than a dozen factors and attempt to eliminate or minimize the effects of each. But the situation is not as complicated as it might at first sound.

A number of the factors relevant to osteoporosis are medical, surgical, and therapeutic conditions of which every physician should be aware. These include abnormal function of the thyroid and parathyroid glands and the chronic use of

---

Until a means to prevent osteoporosis is discovered,
everyone should insure an adequate intake of
calcium...[and]...everyone should engage
in a regular program of exercise.

---

anticonvulsants and steroidal antiinflammatory drugs. Osteoporosis is but one of many reasons to avoid smoking and alcoholism. Heredity may be important: Whites and Asians are more likely to be affected; hip or vertebral fractures in your parents, grandparents, or other relatives put you at increased risk. Female sex hormones are certainly of importance, at least in the short term; surgical removal of the ovaries brings on the same acceleration in the rate of bone loss as does normal menopause. For reasons that remain obscure, women who have never borne children are more likely to experience osteoporosis.

Until a means to prevent osteoporosis is discovered, every woman should discuss estrogen replacement with her physician at the time of menopause. For those with a strong family history of osteoporosis or other risk factors, the decision for estrogen replacement therapy may well be a wise one. Until a means to prevent osteoporosis is discovered, everyone should insure an adequate intake of calcium. As was discussed at length in Chapter 17, this most emphatically does not mean resorting to calcium pills. Until a means to prevent osteoporosis is discovered, everyone should engage in a regular program of exercise.

In Chapter 19 we saw how muscles respond to inactivity by shrinking and to stress by growing. There is no doubt that bones respond in a similar fashion, and there is good reason to believe that exercise may be an effective way to increase bone mass. Indeed, the idea that exercise may prevent osteoporosis is so attractive that many have accepted it as proven. Though it is my guess that a scientifically designed and evaluated program of exercise will ultimately be shown to be an effective preventive measure, the present evidence for that view is less than overwhelming.

We have so far considered osteoporosis as either "senile," a consequence of aging, or "postmenopausal," a process related to loss of estrogen. Physicians early in this century recognized a third form that they called post-traumatic osteoporosis. It refers to the observation that the inactivity that often accompanies injury or disease results in bone loss.

The major forces that act upon bone are muscle tension and gravity. Victims of paralytic polio suffer bone loss along with atrophy of their muscles. Astronauts in space, freed of the forces of gravity, invariably begin to excrete more calcium than they take in. As unnatural as we might wish to consider the weightlessness of space or muscle paralysis, the response of bone to these situations is entirely normal. Despite the presence of calcium in abundance, bone will not be formed in the absence of stress. It comes as no surprise then that loss of bone from inactivity cannot be prevented by calcium supplements. If lack of stress on bones causes them to become thinner than normal, might not the imposition of greater than normal stress cause them to become thicker? That reasonable question and its plausible answer are the basis for our hope that exercise may prevent osteoporosis.

It is hardly a new idea that lack of exercise may be an important risk factor for osteoporosis. I've already told you how A. R. P. Walker's finding in the 1950s that Black South Africans suffer only one-tenth the number of hip fractures as Caucasians led him to dismiss dietary calcium as a significant element in osteoporosis. Noting the greater physical demands placed on the Bantu compared with Whites, Walker also suggested that the role of exercise had probably been underestimated. Might it be that the racial differences in the incidence of osteoporosis that have so consistently been observed around the world owe rather less to subtle genetic influences and rather more to the activities of everyday life?

Might our goal of giving a white woman the skeleton of a black man be a matter for the exercise physiologist rather than the genetic engineer?

A simple way to assess the effects of exercise on bone density is to compare athletes with non-athletes. An investigation by Drs. Bo Nilsson and Nils Westlin of the University of Lund in Sweden was the first of its kind, and it has served as a model for many similar studies. Their subjects were 64 athletes and, as controls, 39 "healthy men." The density of the thigh bone was measured in each subject. The athletes included weightlifters, runners, soccer players, and swimmers; nine had represented Sweden in international competition. The control subjects were divided into those who exercised regularly and those who did not.

The results obtained by Nilsson and Westlin certainly support the idea that exercise increases bone density. As a group, the international-class athletes had a density half again as great as the non-exercising controls. However, a far more important observation concerned the effect of exercise on the "non-athletic healthy men." Ordinary men who exercised had a 21% greater bone density than those who did not and their densities were only 18% less than the most elite of the athletes, those who engaged in international competition. Indeed, there were no statistically significant differences between the exercising non-athletes and the national-class runners, soccer players, and swimmers.

The study by Nilsson and Westlin was published in 1971. Those of us taught by television to believe that medical science proceeds via a sequence of "breakthroughs" may think 1971 to be a long time ago. In fact, though we continue to be encouraged in our belief that exercise may help to prevent osteoporosis, there are a number of reasons why we cannot regard the case as proven.

Those investigations that have compared very active people, such as marathon runners, with sedentary individuals have found modest differences in body calcium and bone

density. The problem with all such studies is that we seldom have the data to decide whether the marathoners and
dedicated athletes had more dense skeletons to begin with.
Might some unknown factor cause both increased bone
density and athletic excellence or the desire to run 26 miles
at a time? In fact, several quite recent studies, including one
by Bo Nilsson and Olof Johnell, have compared women who
have suffered fractures with those who have not and have
failed to identify physical activity as a risk factor.

Other problems plaguing those who would prove that
exercise prevents osteoporosis include the site at which bone
density is measured. In the past this has often been at the
distal radius, a site at which fractures frequently occur in
elderly women. However, there is no good reason to believe
that distal radius density is related to that in the hip or vertebrae, fracture sites of far greater concern. The physical means
of measuring bone density have improved greatly in just the
last few years and there is concern about the validity of some
earlier studies. Definitions and quantitation of physical
activity are seldom as precise as one would like; Nilson and
Westlin in their admirable study simply asked the control
subjects whether they regularly exercised.

I think I've said just about all the negative things I can
about exercise. The fact is that I think exercise presents the
best hope for the prevention of osteoporosis. I expect this to
be established beyond question within this decade, and we'll
now consider two studies supporting that contention.

The first was done in the early 1980s in Denmark by Bjorn
Krolner and his colleagues. It differed from all the other
studies we have so far considered in that it was prospective
in nature. Instead of comparing a healthy active group such
as marathoners with a sedentary one, the Danes studied
women of average age 61 who already had suffered an osteoporetic fracture of the arm. The women were divided into
an exercise group, which engaged in a varied program for
an hour twice a week, and a control group, which maintained

> *If exercise is good for you in terms of decreasing*
> *the risk of osteoporosis, then the more the exercise*
> *the less the risk....Wrong....With respect to the health*
> *of our bones, Bloomingdale's Law must be changed:*
> *you cannot be too rich, but you can be too thin.*

their accustomed lifestyle. Over the 8-month course of observation, the mineral content of the lumbar spine in the control group decreased by 2.7%. In the exercise group, there was an increase of 3.5%. Though granting it remained to be established that such bone density changes would be reflected in a decreased incidence of vertebral fractures, the authors concluded that exercise may prevent spinal osteoporosis.

A similar conclusion was reached with respect to hip fractures by Nicholas Pocock and his associates at the Garvan Institute of Medical Research in Sydney, Australia. In the September 1986 issue of the *Journal of Clinical Investigation* they presented evidence that habitual physical activity increases bone mass in the neck of the femur, the site at which osteoporetic fracture carries the most serious consequences. The study included 84 white women with an age range of 20–75 years, one-half of whom were postmenopausal. An interesting feature of the methodology was that instead of using a questionnaire to assess physical activity, they used oxygen uptake as an objective and quantitative measure of fitness.

If exercise is good for you in terms of decreasing the risk of osteoporosis, then the more the exercise the less the risk. Right? Wrong. The reasons for this are rather roundabout, but interesting. It seems that long ago nature built into women a mechanism for protecting against pregnancy when times are hard. A million years or so ago the Bureau of Labor Statistics wasn't around to advise of the unemployment rate, so Mother Nature used percentage of body fat as a barometer of hard times. When a woman's body fat falls below a certain level, circulating estrogen decreases and ovulation stops; no ovulation, no pregnancy; no ovulation, no menstruation.

For many women, absence of menstruation may not sound like a bad thing. Unfortunately, the low-fat fall in estrogen is the endocrinologic equivalent of premature menopause, which in turn is a powerful influence on accelerated loss of bone. With respect to the health of our bones, Bloomingdale's Law must be changed: you cannot be too rich, but you can be too thin. (I say "our bones" because there is some evidence that a similar phenomenon of low-fat bone loss occurs in men as well.)

Two factors have brought low-fat amenorrhea to our attention. First, many women have become ultrathin as a result of excessive dieting, excessive exercise, or both. Second, menstruation and its associated problems have enjoyed constant media concern. There is not a single issue of a single woman's magazine on the market today that does not contain an advertisement or a feature story on the subject. The magazines devoted to "fitness" are not far behind. They provide in exhaustive detail not only the training schedules of world class women athletes, but their menstrual histories as well. An olympic marathoner tells of three cycles in six years and the women's cross-country team of a West Coast university is reputed to be menstruation-free for a decade.

Miriam Nelson and her colleagues at the Human Nutrition Research Center on Aging at Tufts University have studied in great detail the interactions of exercise with menstruation and bone density in healthy young women. In June of 1986, they described 11 women runners from the Boston area who had not menstruated for at least a year (the average was more than 2 years). When compared with 17 normally menstruating runners, the amenorrheic group was found to have a lower bone density in the lumbar spine. A simple explanation for this finding was that the nonmenstruating women also had only about one-third as much estrogen circulating in their blood. Contrary to expectations, the normal and amenorrheic women did not differ in their miles run per week or percent body fat. Indeed, the normally menstruating women were slightly thinner (19.7% versus 21.6% body fat) and ran slightly more (40 miles per week versus 35).

If exercise increases bone density and excess loss of body fat indirectly leads to a decrease in bone density, where is the appropriate balance point; how does a woman know when she's gone as far as she should go? Some have suggested that it is the amount of exercise, perhaps miles run per week; others that it is the percentage of body fat. Dr. Nelson's results indicate that neither is a completely reliable guide. Instead, the best indicator may be menstrual irregularities or cessation of menses. Most exercising, dieting women will never reach that point. Those who do may wish to carefully balance the benefits of high level training or an unusual degree of thinness against possible harm to their bones.

I earlier said that I expect the preventive effects of exercise on osteoporosis to be essentially proven in this decade. That perhaps too optimistic prediction will come true if we can make rapid progress in identifying specific means of exercise that will reliably increase bone density in those areas, especially hip and vertebrae, which are subject to osteoporetic fracture. At the present time the vague recommendation of "weight-bearing exercise" is universally made. I don't disagree with that suggestion; any exercise that gets you on your feet and moving is a good idea. But we must and should be able to become far more specific in our exercise prescription. For example, might we learn how to harness the remarkable weight-training technology, now devoted to the largely narcissistic task of building muscle, for the specific purpose of building bone?

Finally, let's return to the landmark 1971 report by Nilsson and Westlin for a word about swimming. Their statement that "the swimmers did not differ significantly from the controls" has often been repeated by others and has caused many to reject swimming as a useful exercise with respect to osteoporosis. This is unfortunate. A closer examination of the data reveals that the swimmers did not differ from the controls who exercised. In fact, exercising-controls, swimmers, runners, and soccer players were statistically indistinguishable from one another, and all had more dense bones than the non-exercising controls. All those who are able should walk, run, dance, and otherwise "bear weight." But for those who can't, keep on swimming; your bones as well as your heart will appreciate it.

Summary

Over the past half-century, various "causes" and various "cures" for osteoporosis have been proposed; none have stood the test of time. We are today as uncertain as ever about means to prevent the enormous financial and human cost of the disease. Today's preventive approach is essentially three-pronged: (1) adequate calcium intake during the period of maximal skeletal growth; (2) regular exercise throughout life at an intensity compatible with regular menstruation; and (3) estrogen replacement therapy [ERT] after menopause. The first two of these are quite reasonable. The third, ERT, remains a matter of considerable debate; some appear to believe it entirely safe; others continue to worry about possible promotion of cancer and other adverse effects. Perhaps it is time for the National Organization of Women or some similar group to bring pressure on the nation's predominantly male Congress for a much accelerated program of research. We may hope that sometime in the future molecular biology will provide a completely successful, perfectly safe way to stimulate the growth of bone.

CHAPTER 25

The Beverly Hills Diet

and Other Tales of the Supernatural

Sent to study planet earth, an extraterrestrial being might reach an odd conclusion about the diseases that afflict the citizens of the United States.... analysis of our newspapers, magazines, radio, and television...would suggest that our major concern is body weight.

Sent to study planet earth, an extraterrestrial being might reach an odd conclusion about the diseases that afflict the citizens of the United States. To be sure, cancer and heart disease kill most, and AIDS, though totally preventable, causes great anxiety. But analysis of our newspapers, magazines, radio, and television, together with a monitoring of the thoughts of American men and women, would suggest that our major concern is body weight. In toting up the numbers, our visitor would find that, at any given moment, three-quarters of the female members of the population are too heavy and are "on a diet." The males are a little different; up until the age of 25 or so, many actively try to gain weight, but then they too begin to diet, though seldom with the ferocity of the female of the species.

How might we explain to our visitor this preoccupation with body weight, more specifically, this passion to remove fat from our frames? Is it mere fashion? We must hope not; fashion has always been a fraud. Is it health? Americans certainly are concerned about their health, but as we grow

---

Can you or I, as individuals, take guidance
from tables of ideal weights? Indeed, is there
such a thing as an ideal weight?

---

healthier as a nation we seem to feel less well as individuals. Let's begin by considering the notions that body weight influences health, that there is a disease called obesity, and that for each of us there is an ideal weight.

In many places at many times throughout human history, fatness was a goal to be achieved, a status symbol. The most successful members of a society, those who had more than enough to eat, didn't drive Mercedes or belong to country clubs or ski in Aspen; they grew fat. (Oh, how some of us long for those simpler days.) Several factors have led us from fat as a status symbol to a general desire for thinness. First of all, a status symbol that nearly everyone can have stops being a status symbol. In a society such as ours, where even the poorest can be fat and often are, fatness now more often is seen as a lack of discipline than of prosperity. Thinness also has a medical rationale; it is said that being fat is unhealthy; we'll consider that idea first.

Insurance companies have a vital interest in knowing how long people, on average, can be expected to live and have known for some time that very fat people tend to die sooner than those of average weight. In October of 1942, Louis Dublin, Chief Actuary of the Metropolitan Life Insurance Co., formalized the relationship between fat and longevity by publishing a "Proposed Range of Ideal Weights for Women," and soon followed with a similar table for men. With the passing years, the Metropolitan tables have gone through several revisions, but they remain a point of reference for all who would lose or gain weight.

An actuary is one who applies to matters of insurance the methods of statistics. Mark Twain attributed to Disraeli the remark that "There are three kinds of lies: lies, damned lies, and statistics." Can you or I, as individuals, take guidance from tables of ideal weights? Indeed, is there such a thing as an ideal weight?

Consider two groups of athletes: marathoners and football players. The top ten finishers among the men in the most recent United States Olympic marathon trials averaged 5 feet 10 inches in height with an average weight of 147 pounds. That puts them at the low end of the acceptable range in the 1983 revision of the Metropolitan tables. In contrast, the 1990 Super Bowl Champion San Francisco 49ers averaged 231 pounds, too heavy according to the table. The typical interior lineman was 6 feet 4 inches tall and weighed 273 pounds; that's at least 65 pounds too much if the Metropolitan guide is followed.

Are we to conclude that marathoners are thin, but healthy whereas professional football players deviate so far from their "ideal weights" as to be dangerously overweight? The answer is a most emphatic "no." Instead we must conclude that tables of recommended or ideal weights, whatever their actuarial value, are without merit when applied to individuals. This should come as no surprise; in 1942, Louis Dublin himself said "there is no one set of weights that can be called 'ideal' and to which all...of a given height should conform."

But if there is no such thing as an ideal weight, how are we to distinguish between the healthy overweight (professional football players, for example) and those who are morbidly obese? There are several ways to answer that question. Among the less satisfactory is to say that obesity, a disease, is present when body weight exceeds by some percentage that of the average person; many use a figure of 20%, others as little as 10%, in excess of the average. A further refinement is to express weight and height with a single number called the

...tables of recommended or ideal weights...
are without merit when applied to individuals.

...it is not our body weight, but our body fat that should concern us...[and]...even in the presence of excess body fat, we must consider the consequences of removing that excess...before launching an all-out war upon it.

body mass index (BMI or Quetelet index, defined as weight in kilograms divided by the square of the height in meters). As far as our interior linemen are concerned, it matters not which criterion we choose. At 6'4" and 275 lbs, they are 33% over the recommended Metropolitan weight and have a BMI of 33.49, well in excess of the usual cutoff for obesity at 30.00.

There are in fact two keys to a rational approach to body weight. First, it is not our body weight, but our body fat that should concern us. Second, even in the presence of excess body fat, we must consider the consequences of removing that excess from the individual before launching an all-out war upon it.

Scientists and physicians have a number ways to estimate body fat; calipers applied to skin folds at various sites, the determination of body density by total immersion in water, and the use of various radioactive tracer dilution methods are examples. Unless you already find yourself in the hands of such professionals, I suggest that you forget about them. The vast majority of us needs nothing more than a full-length mirror and our naked bodies. Obvious rolls of fat are unequivocal evidence of excess weight. Justice Potter Stewart's definition of pornography fits obesity as well: "I can't tell you what it is, but I know it when I see it."

As a scientist, I like to confirm my visual assessment in a more objective manner; my scientific device is called the "reference trousers." The trousers come from a time when I was much younger and quite happy with my body. Whatever else the passing years may do to me, so long as I can fit into the reference trousers and so long as my full-length mirror image isn't too troubling, I'm doing all right with respect to body fat. It turns out that the reference trousers and the mirror tell me the same thing: At age 50, I should weigh about 10 pounds less than I did as an athletic 21 year old. (Depending upon where your excess fat tends to settle, you may wish to use some other item of clothing as your reference; just remember that it should have fit well at a time when you liked your body, not when you were the fattest of your life or 8-1/2 months pregnant.)

Let us imagine that you have failed both the mirror and the reference trousers tests. There are two reasons why you may wish to lose weight: (1) matters of health and (2) matters of appearance. To settle the first you need simply visit your personal physician for a routine physical examination. It needn't be one of the fancy, $500 kind that is so attractive to some doctors and their accountants, but it should include a careful assessment of your medical history and your current physical state. Even people who fall well within an "ideal weight" range may benefit from a further loss in body fat when conditions such as diabetes, high blood pressure, or hyperlipidemia are present, which is still another reason not to depend on an actuary for advice on body weight. Of course, a medical reason may be the starting point for your desire to lose weight. We discussed this at length in Chapter 23.

Those who attempt to lose weight for well-defined reasons of health are possibly the most successful of all dieters; you just can't beat a clear-cut incentive. But let us now imagine that, whatever the mirror and the trousers say to you, your doctor finds no specific medical reason to require weight loss. You may of course fall back on statistics; on average and excluding extremely low body weight, the thinner you are the longer you will live. But as attractive as a long life is in the abstract, it's not much of an incentive as you bring your ravenous appetite to dinner. Instead, most of us diet in an attempt to improve our appearance. There's nothing wrong with that so long as our goals are realistic; for a woman born

---

...as attractive as a long life is in the abstract....
most of us diet...to improve our appearance. There's
nothing wrong with that so long as our goals are realistic;
for a woman born with the genes of a Russian shotputter
to covet the jeans of a Bolshoi ballerina
is to invite frustration and failure.

---

with the genes of a Russian shotputter to covet the jeans of a
Bolshoi ballerina is to invite frustration and failure.

To lose weight one simply needs to decrease energy intake below energy expenditure. To make up the difference,
the body will convert glycogen and protein and fat to energy; weight will be lost. If the principle is so elementary why
do so many have so much trouble doing it? The answer to
that simple question is equally simple: pleasure—food is fun.
Compare sex, a great driving force in human affairs. In terms
of urgency, longevity, and cumulative satisfaction, sex pales
in comparison with food as a reinforcer. Because few wish
to diminish the total pleasure of their lives, successful weight
loss requires that we find reinforcers to substitute for food.

If the reason I wish to lose weight is a lack of satisfaction
with my appearance, it would seem an easy matter to balance a
decrease in "food-pleasure" with an increase in "appearance-
pleasure." The difficulty in making that exchange has a lot
to do with timing. Food provides instantaneous satisfaction;
loss of body fat takes time, a change in appearance even
longer. Whatever I may tell myself will happen in the long
run, the simple fact is that saying no to a three-scoop sundae
with nuts, hot fudge, whipped cream, and a cherry on top
denies me pleasure right now, but when I look in the mirror
tonight my appearance will be unchanged. It is this lack of
immediate reinforcement in the form of an improved
appearance that makes weight loss so difficult; pleasure has
been lost with, quite literally, nothing to show for it.

In the remainder of this chapter we will consider the physiological principles of energy storage and release. We will examine differences between individuals; differences that allow some to maintain a slim body without effort while others appear doomed to fatness. I will also tell you about a number of different "diets," their underlying rationale, and whether they are likely to be of value to you. We will examine the role of drugs and of artificial sweeteners. In none of this discussion should we lose sight of two simple facts: (1) Food and the rituals of eating bring pleasure, human pleasure of the most fundamental kind. (2) Successful weight loss requires, to a greater or lesser degree, the denial of food-associated pleasure.

As was pointed out in Chapter 2, an obvious function of body fat is as a means to store energy. The most common unit of measurement of that energy is the calorie, the amount of heat required to raise the temperature of 1 kilogram of water by 1 degree centigrade. Fat is a very efficient form of energy storage; it provides 9 calories for each gram or about 250 calories per ounce. This is more than twice the energy content of either protein or carbohydrate. Though it is an oversimplification to imagine that excess energy intake is converted to body fat at the rate of 9 calories per gram, and it is an oversimplification to imagine that body fat is lost at a rate of 1 gram for each 9-calorie deficit, that 9 calorie per gram figure does provide a useful point of reference.

In the first three chapters of this book, we considered the major sources of food energy, the so-called macronutrients. We learned that animal and vegetable proteins are broken down into smaller units, absorbed from the gut, and then reassembled as human protein. Likewise, fat and carbohy-

Because few wish to diminish the total pleasure of their lives, successful weight loss requires that we find reinforcers to substitute for food.

---

*A human body subjected to a diet deficient
in energy first turns to carbohydrate
and only grudgingly gives up its fat.*

---

drate in what we eat become fat and carbohydrate of our bodies. It should then come as no surprise that when I go on a diet, i.e., when my food intake can't meet my energy needs, my body begins to consume itself; energy stored as fat and carbohydrate (glycogen) and protein is used to make up my energy deficit. "But," my mind protests, "I don't want to lose protein and glycogen; I want to lose fat." "Too bad," my body replies, "I make these decisions."

A human body subjected to a diet deficient in energy first turns to carbohydrate and only grudgingly gives up its fat. Oddly enough, this phenomenon of nature has led to more than one fortune being made in the weight-loss business. Because of the respective energy content of fat and of glycogen and because of their forms of storage, many otherwise idiotic "diets" appear to work very well in the short run. Let's consider some numbers.

Imagine that I am a bit pudgy, but stable in weight on a diet providing 3000 calories per day. In an attempt to reduce, I decrease my caloric intake by 1000 calories. My body will immediately make up the difference by breaking down glycogen into glucose. At 4 calories per gram, that 1000 calories should lead to a daily weight loss of about half a pound. Not bad but, in fact, it's even better than that. For each gram of glycogen converted to glucose, three grams of water are released and excreted. Now the 1000 calories convert to a weight loss of over two pounds.

Were I to continue to burn up my 12 pounds or so of stored glycogen at the rate of 1000 calories per day, I would lose, in a mere three weeks, nearly 50 pounds! That's the good news. The bad news comes in two parts. The first is that I haven't lost any fat. "Miracle Diets" that promise rapid weight reduction invariably depend upon glycogen depletion with its accompanying loss of body water. The second piece of bad news is more fundamental. Despite the unassailable logic and arithmetic by which I projected a weight loss of 50 lbs in 3 weeks by a caloric deficit of 1000 calories per day, human bodies simply don't work that way.

Instead of a constant using up of glycogen, alternative sources of energy will be sought; stored protein and fat are these sources. Because fat provides more than twice the amount of energy per gram as does glycogen and because fat depots contain very little water, the rate of weight loss slows. In addition, the body becomes more efficient in its use of energy. The net result of these changes is that after a short period of dieting, weight loss is not at the rate of 2 pounds per 1000 calorie deficit, but perhaps one-tenth as much. It is sometime during this shift from glycogen-burning inefficiency and quick results to the jealous guarding of body fat that most dieters founder. The "quick weight loss plan" that seemed so exciting and so effective (and often so expensive) is soon discarded. Lost weight is restored and we again await the next "nutrition breakthrough" with its promise of slimness without effort.

There are a variety of ways to counter these rather dismal facts of metabolic life. Far and away our most important weapon is exercise, both aerobic and anaerobic. But before looking at the role of exercise in weight loss, let's consider what I earlier called "the differences between us": the reasons why some people need never diet, while others must relentlessly pursue weight loss with generally unrequited passion.

It is undeniable that social, psychological, and cultural factors influence the incidence of overweight. For example, in this country, obesity is six times as prevalent among poor women; just the opposite of what would be predicted if ability to buy food determined the degree of overweight; a simple explanation is that visible fat is more accepted and acceptable among the poor than among the more affluent.

Genetically determined differences...largely explain why,
when two persons are fed the same number of calories,
one of them may lose weight while the other gains.

But without diminishing the role of factors such as these, we must acknowledge something more fundamental: our genetic endowment. One characteristic of the genes that our parents give us is our basal metabolic rate (BMR). This is a measure of the amount of energy required just to stay alive.

It comes as a surprise to some that in sedentary adults the basal metabolic rate, our minimum energy requirement, accounts for about 75% of total daily energy expenditure. Furthermore, BMR varies significantly between individuals. Genetically determined differences in BMR largely explain why, when two persons are fed the same number of calories, one of them may lose weight while the other gains. It is ironic that a low basal metabolic rate is advantageous to humans in a setting of food scarcity, though predisposing to obesity those for whom food is abundant.

Whatever the differences between individuals with respect to BMR, weight loss still would be a rather straightforward matter if each of us could plan on having the same basal rate of energy use throughout our adult lives. Make no such plans; BMR changes in response to caloric intake. Earlier I mentioned that the decrease in rate of weight loss with sustained dieting is partly a result of the body's "more efficient use of energy"; a decrease in BMR largely accounts for that effect. A study conducted in The Netherlands by Dr. Janna de Boer and her colleagues is illustrative.

Fourteen women with an average weight of 206 pounds were put on a 1000 calorie per day diet for 8 weeks; their average weight loss was 22 lbs. However, over the course of the 8-week diet, basal metabolic rate decreased by an average of 15%. The practical consequence of the change in BMR is that these women would have to continue indefinitely to eat 15% less than before simply to maintain their weight loss. As many readers can attest, it is a most frustrating experience for those who have managed to lose a few pounds to find that they cannot return to their old eating habits without quickly returning to their old weight.

The fact that basal metabolic rate plays an important role in determining energy needs together with the observation that the rate can change in response to an excess or deficiency of food have led to something called "the set point theory." The thought is that perhaps body weight is regulated in the same way as is body temperature or the composition of the blood. Thus, each of us would have a predetermined "natural" body weight. Movements above or below the set point would call into action compensatory mechanisms, changes in basal metabolic rate, for example. Those we call obese would simply be those whose natural set point clashes with what current fashion calls attractive or contemporary medicine thinks to be healthy.

Some, including a few overweight professionals, have used set point theory as an excuse not to lose weight; they say that some of us are "naturally" obese and some are "naturally" lean, and there's nothing to be done about it. The weight of scientific evidence does not support this position. Though we may accept as fact the existence of physiological mechanisms that function to conserve body fat, and though we may also accept as fact that there are individual differences in the level of body fat that triggers those mechanisms, there is no reason to conclude that weight loss is impossible for anyone. Rather than acceptance of overweight as our genetically determined destiny, we should use our knowledge of how the body defends against loss of fat as a further incentive to the practice of patience and discipline, virtues required for life-long weight control.

In this land of abundance, the probability of unaided, permanent weight loss by a moderately obese person is about

---

Of particular importance is the effect of exercise upon the form of weight loss that occurs. Exercise favors the depletion of fat and helps to maintain muscle protein.

---

the same as spontaneous 5-year abstinence by a teenage heroin addict. The prospect of an unaided battle against a relentless "set point" is grim indeed. The key word is "unaided"; by it I mean "without thought," "without a plan," "without resort to the social, psychological, and medical research of the last fifty years." Now you may think that I am about to tell you of a "nutrition breakthrough," a "startling scientific advance," a "secret known previously only to an ancient South American tribe" and recently rediscovered by one of Europe's "top docs," a way of "flushing away fat while you sleep," or an "effortless and guaranteed program of weight loss while eating all that you want." I have no such intention. Instead, I offer a few simple principles that, diligently applied, promise a better body.

The value of regular aerobic exercise cannot be overemphasized. In Chapter 20, we examined some simple ways to gain a training effect and in Chapter 23 the role of exercise in reducing the risk of heart attack and stroke. Now let's consider the diverse ways in which exercise facilitates the achievement of your ideal weight.

The most obvious of the effects of regular exercise is that it requires energy. For every calorie used in exercise, one less calorie need be cut from the diet. Doomsayers like to point out how little energy over and above the basal metabolic rate is required by even the most strenuous forms of exercise. For example, in jogging one mile I increase my energy use by only about 120 calories; that's about the number in 10 ounces of Coke. But multiply 120 calories per mile by 10 or 15 or 20 or more miles per week, week in and week out, and the energy use becomes rather impressive. As a bonus, there is evidence to suggest that exercise decreases rather than increases appetite, and that the decrease in BMR caused by dieting is at least partially offset by exercise.

Of particular importance is the effect of exercise on the form of weight loss that occurs. Exercise favors the depletion of fat and helps to maintain muscle protein. As the sellers of packaged foods have become more bold in making health claims, you may have seen advertisements claiming that eating a lot of protein will help you maintain muscle while losing weight. If there be such an effect, it must be small relative to what exercise can do. Many who follow training programs such as that outlined in Chapter 20 find that their body weight does not change very much, but that their appearance does; this is especially likely if a modest amount of weight training is done as well; muscle is more dense than fat. Our "naked before the mirror" and "trousers tests" are far better indicators of a healthy body weight than any rigidly prescribed "ideal weight."

In addition to the well-defined physical effects of exercise in fat-control, there are other benefits as well. Most important of these is a changed mental state. I am not speaking of an orgasmic "runner's high" that few have ever, or will ever, experience. Instead, I refer to a general sense of well-being that is universal among regular exercisers. It matters little that we don't presently know the precise chemical origin of this change in state. What is important is that dieting becomes easier. Exercise is not only a diversion from eating and a consumer of calories, it becomes a substitute reinforcer as well. It is for this reason that I cannot conceive of a lifelong program of weight control and fat loss that does not include a component of regular exercise.

Now let's consider diets. Each of the thousands of weight-loss schemes that have been presented to the public

---

...there are other benefits [to exercise]. Most important of these is a changed mental state....I refer to a general sense of well-being that is universal among regular exercisers.

---

may be classified with respect to two factors: balance and caloric content. Balance refers to the relative proportions of the macronutrients (fat, protein, and carbohydrate) and to the content of essential vitamins, minerals, and fatty acids. As we saw in Chapter 22 there is currently much controversy regarding what constitutes ideal balance for the macronutrients. However, any plan that virtually eliminates one of them, or an essential micronutrient, might reasonably be called "unbalanced." Caloric content has to do with the amount of food energy provided by the diet; the degree of caloric restriction may vary from 0 to 100%.

It is easy to imagine a diet in which caloric restriction is 100%; we usually call it fasting or, more bluntly, starvation. Fasting has a long history in every major religion of both the Eastern and Western worlds. When applied to weight loss rather than to the enhancement of spirituality, starvation would appear to be simple, direct, and nearly ideal; it is none of these. Aside from the fact that few of us are able, by force of will alone, to give up all food, starvation may have long-term adverse effects.

Much of the confusion about diets and weight loss arises from the popularization of what are essentially experimental procedures in massively obese persons. Typically, a dietary plan is evaluated on a metabolic ward of a research hospital and the results published in the medical literature. Democracy then takes over—I am free to read the research report, interpret it as I wish, go on television, write a book, preach on any street corner. The scientific merit of the study and my qualifications to interpret it are irrelevant; I am merely exercising the right to free speech guaranteed by the Constitution. "Therapeutic starvation" and its derivative, the "modified fast," provide excellent examples of this phenomenon.

Extensive descriptions of starvation as a medical treatment first appeared early in this century. Targets for the treatment included not only obesity, but also diseases of the skin, multiple sclerosis, cancer, and a variety of other disorders for which no fully satisfactory therapy was available. However, the modern era of starvation as a means of weight loss began with the appearance in 1959 of a paper by Walter Lyon Bloom, MD, entitled "Fasting as an Introduction to the Treatment of Obesity." In it he described the effects of fasting for short periods of time. A total of nine patients participated. The four men weighed an average of 262 pounds, the five women 250 pounds. All had tried various diets before; all had lost and regained weight.

Weight loss in Dr. Bloom's patients occurred about as we would predict and averaged 2.5 lbs per day. More remarkable was the observation that the patients experienced "little hunger in the sense of a compelling drive to satisfy a need" and all had "a sense of well being." Furthermore, "an unexpected result of the study was the patients' satisfaction with a 600–800 calorie diet following the completion of the fast, whereas prior to the fast this type of diet had been very unsatisfactory."

Before rushing to begin your fast, I must tell you that not all have observed the same effects as did Dr. Bloom. For example, many who have undergone forced starvation have been tormented by unending thoughts of food. The suggestion that a short period of fasting thereafter promotes adherence to a 600–800 calorie diet must be a rare phenomenon; were it true in general, there would be no involuntarily obese among us. Nonetheless, many physicians were led by Bloom's results to try therapeutic starvation in the treatment of obesity; at least one of the consequences was totally unexpected.

There is no question that starvation will produce very considerable weight loss. But, as I mentioned earlier, all that is lost is not fat. Carbohydrate in the form of glycogen is rapidly depleted, but even then the body does not shift to burning fat alone. During a 3-week period without food, about a third of the total weight loss will be in the form of protein. As a result, all muscles of the body will be diminished in size and strength—and the heart is a muscle.

In Dr. Bloom's study, the maximum period of starvation was 9 days. In the decade that followed, others went much further. A few patients died during fasts of several weeks, but little attention was paid; death was attributed to diabetes or hypertension or heart disease present before the fast was begun. However, in May of 1969, Dr. E. S. Garnett and his colleagues of Southampton (England) General Hospital reported a case that was not easily dismissed. "An obese but otherwise healthy 20-year-old girl" died after 30 weeks of starvation and after having reached her "ideal weight." During the fast she had received vitamin and mineral supplements, together with essential amino acids. The immediate cause of death was ventricular fibrillation; autopsy revealed widespread destruction of muscle fibers of the heart.

The obvious drawback of starvation as a means of losing weight is that our bodies simply refuse to burn only fat; protein is lost as well and when protein is lost from the heart, disaster may strike. Why not replace the protein? The simplicity and plausability of that idea are so attractive it was soon tested in France, England, Germany, and the United States; in 1973, the idea was named "the protein-sparing modified fast." Instead of pure starvation, obese patients were fed 400–800 calories per day as protein. Unfortunately, before such diets could adequately be investigated in either humans or animals, an osteopath named Robert Linn exercised his constitutional rights and wrote a book called *The Last Chance Diet*.

Published in July of 1976, Dr. Linn's book became a best seller. Subtitled "—When everything else has failed," "Dr. Linn's Protein-Sparing Fast Program" promised immediate and rapid weight loss without hunger and, in Dr. Linn's words, "no major side effects." The heart of the diet was a substance modestly called "Prolinn," a predigested protein from beef hides. In addition to 600 calories per day in the form of the beef-hide extract, the dieter was to take multivitamins, calcium, folic acid, and potassium, together with two quarts of water, tea, coffee, or a diet soft-drink.

Within 18 months, about 100,000 Americans tried *The Last Chance Diet* using either Prolinn or one of some 50 other liquid protein preparations that quickly appeared in drugstores, supermarkets, and health food outlets. As foods, they were not subject to governmental regulation other than with respect to their freedom from bacterial and other contamination. By the end of 1977, 58 people had been reported to have died while on liquid protein diets. The Centers for Disease Control investigated each and attributed 17 to direct effects of the diet. On December 19, 1977 the CDC and the FDA released details of ten of these deaths. By the end of 1978, horsehide protein hydrolysates were history.

But, a perennial dieter, like Nature, abhors a vacuum.

A quick-fix for the cowhide diet of Linn was to provide "high-quality protein," i.e., an appropriate mix of all the essential amino acids. A more controversial suggestion was to add a small amount of carbohydrate to an otherwise purely protein regimen. The goal was to maintain rapid weight loss, but with less nausea, fatigue, and other adverse effects. One who advocated this approach was Alan N. Howard, PhD. In the early 1970s, Dr. Howard and his colleagues began to investigate various protein/carbohydrate mixes in hospitalized obese persons. Their studies were conducted first at West Middlesex Hospital in London and later at Addenbrooke's Hospital in Cambridge. By the end of the decade of the 70s, *The Cambridge Diet* was ready for sale to the general public.

I have never met Alan Howard, but his medical papers suggest he is a highly competent nutritional scientist, well qualified to design diets for the obese. The original Cambridge Diet provided 330 calories per day—31 grams of high quality protein, 44 grams of simple carbohydrates in the form of sugar, and, as a source of essential fatty acids, 2 grams of polyunsaturated fat, either safflower or corn oil. The macronutrients were supplemented with vitamins and minerals and topped off with a variety of synthetic flavors. Thus the Cambridge Diet is completely balanced, but severely restricted calorically, characteristics intended to allow weight loss at a rate equivalent to total starvation, but without any of the hazards.

As one component of a very low calorie treatment program to be administered to the severely obese under medical supervision, Dr. Howard's diet cannot be faulted. However, when he sold his US and Canadian patent rights to Jack and Eileen Feather, Dr. Howard made the best of decisions and the worst of decisions. Financial success was assured, at least for some; the Feather's were proven hucksters; their earlier creations included Slim-Skins, Trim-Jeans, Sauna Belt, Astro-Trimmer, and, best known of all, the Mark Eden Bust Developer. Each of these products had been offered to the public in advertisements worthy of P. T. Barnum. Dr. Howard's diet would be promoted in the same way but better.

At the heart of the Feathers' marketing strategy were the "Cambridge Counselors." Eileen Feather, billed as "Founder of the Cambridge Plan International," described her counselors as "dedicated and committed people, who want to share with others—to share with you—the marvelous, the fabulous secrets of Cambridge Diet—to welcome you into Our World of science and service so your success is assured, where you simply cannot fail." (Failure is an event well known to the overweight and most plans promise immunity from it. In describing *The Last Chance Diet*, Dr. Linn gave it a new twist: "The program cannot fail. Only you can fail.")

What qualifications were required of a Cambridge Counselor? Ms. Feather tells us only that all of them "have achieved personal goals with the Cambridge Plan—weight loss, improved health, increased energy and vitality, whatever—they have been there." In fact, anyone who enquired about the diet was invited to become a counselor. Upon completion of a two-hour "training program," counselors were given promotional materials for distribution to their friends and neighbors and permitted to purchase cans of the Cambridge formula for $11.67 that they would then sell to their clients for $18.00.

> *The Holy Grail of every overweight person is a diet that permits unlimited intake of food yet guarantees weight loss.*

Counselors were encouraged to recruit others; in February of 1981 there were 25 Cambridge Counselors; 14 months later they numbered 150,000 with annual sales approaching a half billion dollars. No less an authority than Jack Feather estimated net profits in the first full year of operation to be $100 million dollars. Friday morning, September 2, 1983, Cambridge Plan International filed for bankruptcy. It is unknown how many millions remained with the Feathers.

There are several lessons to be learned from the adventures of Dr. Howard and the Feathers. A few have to do with pyramid schemes and who gets rich and who doesn't. We will confine our attention to the scientific. During its brief stay at the top of the diet world, Dr. Howard's formula was tried for varying lengths of time by about 25 million people in the US alone. Few deaths were attributed to it, suggesting that the provision of all essential nutrients confers reasonable safety on diets that are severely restricted calorically.

Though many continue to believe that all very low calorie diets are inherently dangerous and should not be undertaken without medical supervision, others are more sanguine or, perhaps, more realistic. In commenting upon "microdiets" in general and the Cambridge Diet in particular, J. V. Durnan, Professor of Physiology at the University of Glasgow, acknowledges that "the possible dangers of long term use of very low calorie diets make it advisable to use them for only about one month." However, he continues that "For short-term weight loss of modest amounts of 'fat'...there are virtually no contraindications to using very low calorie diets." I must emphasize that he is speaking only of diets fully balanced in vitamins, minerals, and essential fatty acids.

The Holy Grail of every overweight person is a diet that permits unlimited intake of food yet guarantees weight loss.

Balanced and calorically unrestricted plans in fact have had some success in hospital settings. Unfortunately, all nutrients are presented in the form of a liquid whose taste nearly induces the gag reflex and/or is boring in the extreme. The idea is that when food isn't attractive to eye, nose, or palate, less will be eaten.

Let's redefine the dieter's Holy Grail as a diet that permits unlimited intake of delicious food, yet guarantees weight loss. Many books have promised exactly that, and nearly all have been based on a low intake of carbohydrate and an associated metabolic phenomenon called ketosis. I've already explained how carbohydrate restriction rapidly leads to depletion of stored glycogen. As glycogen is used up, large amounts of water are excreted and there is the illusion of rapid fat loss. If caloric intake cannot match the needs of the body, real fat loss begins as the body converts fat and protein to a utilizable energy source: ketone bodies.

During prolonged starvation, even the brain can come to use ketone bodies for energy. It does so with reluctance and often signals its displeasure with fatigue, nausea, loss of appetite, and a generally lousy feeling. Of course, proponents of low-carbohydrate diets use these side effects of ketone bodies as a selling point and tell you that you'll no longer be interested in food. (I am reminded of *Mad* magazine's "Gross Out Diet" in which interest in food is diminished by proposing the use of such techniques as eating spaghetti with an open can of live worms on the table.)

Fat loss is a function of caloric deficit, the difference between energy intake and energy use. Except in circumstances of extreme caloric restriction, extreme energy expenditure, or both, "quick fat loss" simply doesn't occur, nor should we expect it to. In contrast, short term "quick weight loss" is easy to achieve by inducing ketosis with a low- or no-carbohydrate diet. Examples include "The Drinking Man's Diet" (low carbohydrate plus alcohol; also known as the "Air Force," "Airline Pilot's," or "Astronaut's Diet"), the "Royal

Pennington Diet," and the grandly titled "Calories Don't Count." Robert Atkins gave us *Dr. Atkins Diet Revolution: The High Calorie Way to Stay Thin Forever*. The appeal of these diets with their promises of effortless thinness has been and continues to be enormous. Irwin Stillman wrote a book called *Dr. Stillman's Quick Weight Loss Diet*; about six million copies have been sold in various editions.

Other than the fact that their promoters seem to mislead by blurring the distinction between weight loss and fat loss, is there anything wrong with low carbohydrate diets? Nutritionists of course point out that low carbohydrate means of necessity high fat or high protein, or both. Low carbohydrate means severe restriction of the intake of fruits, vegetables, and cereals. In brief, a low-carbohydrate diet runs counter to every modern tenet of sound nutrition. None but the foolish would choose to stay on a low carbohydrate diet for very long. And that brings us to the real deception of such diets: What happens when carbohydrates are reintroduced? A study done by Dr. S. D. Phinney and his colleagues at the Department of Nutrition and Food Science of the Massachusetts Institute of Technology will illustrate.

The MIT scientists were interested in the physical and metabolic adaptations to chronic ketosis. For a 4-week period, nine men ate a diet high in fat (81% of calories) and adequate with respect to protein (16% of calories), but virtually devoid of carbohydrates. At the beginning, all of the men were of normal weight and sufficient calories were provided for energy needs. Despite this, one of the subjects lost five and a half pounds in the first two days of the ketogenic diet. As a group, the results were less impressive, but after 2 weeks the average weight loss was about 2.5 lbs. Adaptation then occurred and at the end of 4 weeks, six of the men had lost an average of 2 lbs while the other other three gained about a pound and a half relative to their starting weight.

I think we can all agree that, for a variety of reasons, it's not a good idea to stay on a high fat, no-carbohydrate diet for very long. "Who cares?" you ask; "If I can lose 10 pounds in 2 weeks, I'll then go back to eating normally; I can handle any diet for 2 weeks." Let's consider the experience of the MIT group. After 4 weeks on the high-fat diet there was an average weight loss of just over a pound. Reintroduction of carbohydrates led in 3 days time to an average weight gain of 7.5 lbs. The quick weight loss, mostly water, was just as quickly regained with some to spare. Although no one has written a book about it, this pointless cycle has been experienced by millions of people taken in by the facile promises of the low-carbohydrate diets.

Let's return for a moment to Dr. Howard's Cambridge Diet. Whatever the abuses that accompanied its promotion in North America, it was scientifically based, completely balanced, and had undergone at least minimal evaluation in a controlled setting before being turned loose on the public. The same cannot be said for dozens of other diet plans.

To sell a book—and that after all is what most diets are about—one needs to grab the prospective buyer's attention. Some names do well: Dolly Parton, Richard Simmons, Victoria Principal, Elizabeth Taylor, James Coco, or any other sometimes-thin celebrity. If a star is not available, then a place where stars or other affluent types might be found will suffice: Beverly Hills, Scarsdale, Palm Beach, Hilton Head, Southampton, Palm Springs. Because excess fat is so often thought a disease, appeal to medical authority is often made: Dr. Atkins, Dr. Stillman, Dr. Burger; many a title begins with "The Doctor's..." and goes on from there. Once again, names of places may be substituted: Cleveland Clinic, Mayo Clinic, or any other prestigious medical, scientific, or educational institution. Words like "revolutionary," "miracle," "quick," "ultimate," and "breakthrough" are quite common.

A few of the diet books listed above provide sound nutritional advice. Most do not. They are instead what Stephen Newmark and Beverly Williamson have called "novelty diets." A novelty diet is one that relies on "the belief that certain

foods, nutrients, or other substances have unique, previously undiscovered, or magical properties to facilitate weight
loss. Emphasis is frequently placed on eating specific food
combinations." Dr. Newmark and Ms. Williamson cite *The
Beverly Hills Diet* as a classic example.

The Beverly Hills Diet was written by Judy Mazel. For a
time in 1981 it was at the top of the nonfiction best seller
lists. Considering a few of the statements made in the book,
"nonfiction" seems an odd category. For example: "it's not
what you eat or how much, but when and what you combine that counts" and "undigested food becomes stuck in the
body and promotes fat." My confidence in the validity of Ms.
Mazel's "theory of metabolism" is a bit shaken by a press
description of the author as "a formerly obese actress turned
nutritionist." On the other hand, she makes an excellent talk-
show guest and her message is hard to beat: You can lose
weight without effort simply by combining your foods according to her "enzymatic laws."

If we strip away the absolute nonsense with which Ms.
Mazel adorns *The Beverly Hills Diet*, the following facts emerge.
Her plan provides about 1200 calories a day mostly in the
form of carbohydrate. The diet is inadequate with respect to
protein, vitamins, and minerals. If you stick to *The Beverly
Hills Diet* for even a few weeks, you will lose weight. Is the
weight loss caused by the magic of proper food combinations?
It is not. Consider my own candidate: *Dr. Winter's Chocolate
Cake and Ice Cream Diet*. Eat all the cake and ice cream you
want up to a daily limit of 1200 calories. Over a 2-week period,
I guarantee weight and fat loss equal to that of *The Beverly Hills
Diet*. (If not satisfied, return all cake and ice cream to me for
a full refund.) The "secret" to both diets is caloric restriction.

Many volumes could be filled with a detailed critique of
the hundreds of "diets" that have come and gone and will
come again. None can deliver what most promise: fat loss
without effort. If even one of them could live up to that claim,
obesity in all but the perverse would disappear from this

---

*...hundreds of "diets"...have come and gone and
will come again. None can deliver what most
promise: fat loss without effort. If even one of
them could live up to that claim, obesity in all but
the perverse would disappear from this planet.*

---

planet. In the next few paragraphs I will make a few simple
suggestions to help you create your own plan for improved
fitness and weight loss. Only you can provide the will to
implement that plan.

(1) Fitness is the primary goal; fat loss will follow natu-
     rally as improved fitness is achieved. Plan your pro-
     gram of aerobic exercise as I suggested in Chapter 20.
(2) As aerobic fitness improves, you may wish to add a
     modest program of strength training. Freedom from
     injury and improved body tone, rather than bulk, are
     the goals, so emphasize light to moderate weights with
     high repetitions.
(3) Once your training program is firmly in place and
     showing results, evaluate the foods you eat. It's at
     this time that a weight goal may be set. A fortunate
     few will need to lose no weight; the training effect
     alone will be enough. For the rest of us, some degree
     of caloric restriction is necessary to achieve an accept-
     able proportion of body fat. If, like George in Chapter
     23, your present diet is patently horrendous, make
     major changes along the lines I suggested for him. All
     of us should keep in mind the following:
     (a) Build your diet around foods that you normally
         eat. "The Brown Rice and Tepid Water Diet" may
         sound like a wonderful way to lose weight, but the
         novelty of "eating like an Indian holy man" soon
         fades. Oddly enough, some of the most wildly

"successful" diet books require a total change in eating habits and patterns, changes people are unlikely to maintain very long in the real world.

(b) Think like a vegetarian. Let fruits, vegetables, and cereals become a larger part of your meals. Rice, pasta, and potatoes prepared with a minimum of fat are fine.

(c) Evaluate your drinking habits. If there be junk food, alcohol is it. The caloric density is high; the 1.5 ounces of whisky in a mixed drink are calorically equivalent to 9 ounces of skim milk. Aside from the energy it provides, and it is energy intake we are trying to reduce, alcohol is a nutritional zero; its calories are "empty." Finally, a few beers or their equivalent make it so much easier to eat everything in sight; "loss of control" is the fancy phrase for this phenomenon.

(d) Maximize your eating pleasure. Trade boring calories for fun calories. For example, drink skim milk instead of whole milk every day, which will then allow you to enjoy a fat-rich piece of cake from time to time.

(e) Relax. Some would have us believe that food is poison, others that food is medicine. Some would have us micromanage our diets: 17.5 calories worth of this and none of that and lots of something else at every one of our (recommended number) of meals. Ignore the extremists; a mixed diet rich in fruits and vegetables and adequate with respect to protein and fat is, as far as science can tell us today, best for you and all your parts.

(f) Try a few gimmicks of your choice. I realize that "gimmick" is not one of the more distinguished words in our language, but it well describes what I have in mind: a strategem to promote a project. Gimmicks for control of food intake include drinking lots of water before every meal, giving up specific foods, not eating after a certain hour, joining a self-help group, keeping a detailed log of what and when and under what circumstances you eat, and on and on. (A more dignified way to refer to these efforts is as "behavior modification.") The most important thing to keep in mind is that not all gimmicks work for everybody, and gimmicks that do work for you may not work all the time—some we can live with forever, others are good for brief periods only. Again, relax and enjoy yourself.

A gimmick that deserves special mention is the use of low-calorie or no-calorie substitutes for sugar. The commercial success of these chemicals is without question; in 1989 Americans spent $5.5 billion dollars on Diet Coke and Diet Pepsi alone. These and other soft drinks are sweetened with aspartame (NutraSweet), one of three chemicals that deserve our attention; the others are saccharin and sodium cyclamate.

The stories of saccharin and cyclamate have a number of curious parallels. Their sugar-like properties were accidentally discovered by graduate students in chemistry, saccharin in 1879 when Constantin Fahlberg failed to wash his hands before supper, cyclamate in 1937 when Michael Sweda held his cigaret with hands contaminated with the chemical. Both came into commmercial use as sugar substitutes, both cause cancer of the urinary bladder in rats, the banning from human use of both has been called for. Indeed, cyclamates were banned in the United States in 1970 and saccharin has avoided that fate only by act of Congress. Though still available for use, all saccharin-containing products must now carry the warning that it "has been determined to cause cancer in laboratory animals"; not quite the message that an advertising agency might suggest.

Burdened as are cyclamate and saccharin with the stigma of carcinogenicity, the field has been left to aspartame.

Strictly speaking, aspartame is not "nonnutritive"; it has about the same caloric content as does sugar, but because it is 200 times as sweet as sugar on a weight basis, its caloric contribution in sweetened products is negligible. Developed in 1965 by the American pharmaceutical house, G. D. Searle, aspartame is a dipeptide composed of the amino acids, aspartic acid and phenylalanine. The latter is one of the essential amino acids discussed in Chapter 1.

Aspartame, or NutraSweet as Searle calls it, received FDA approval in 1974. Almost immediately, studies appeared indicating that aspartame, in combination with monosodium glutamate (MSG), a flavor enhancer, may cause brain damage in mice. Furthermore, suspicions were aroused that some of the safety data presented to the FDA as part of the approval process might have been falsified by the commercial testing laboratory employed by Searle. As a result of these findings and the need for a complete reevaluation of the chemical, aspartame-sweetened products did not reach consumers until the summer of 1981. Two years later, the FDA gave permission for the use of aspartame in soft drinks.

Searle's patience had finally paid off. In 1989 more than a billion dollars worth of NutraSweet was sold for use in Diet Pepsi, Diet Coke, and about 1500 other "low-cal" products. Nonetheless, in the tradition of saccharin and cyclamate, there has been continuing controversy about its safety. Claims have been made that aspartame causes headache, convulsions, anxiety, and a variety of other disorders both medical and behavioral. In 1988, Hymen Roberts, MD, director of the Palm Beach Institute for Medical Research, suggested that Nutra-Sweet is related to confusion, loss of memory, and the development of Alzheimer's disease. (Dr. Roberts' name may sound familiar; in Chapter 16 I told you of his claim that excess vitamin E causes pulmonary thromboembolism.)

In 1985, the Food and Drug Administration began a program to monitor complaints about adverse reactions to food ingredients. In the first three years of the program, 4800

There are several possible explanations for the paradox
of an ever-fatter nation consuming
ever-increasing amounts of diet foods....
[perhaps]...users of diet drinks, confident that many
calories have been saved, go to excess with other foods.

(80%) of the complaints had to do with aspartame. Nonetheless, officials of the FDA and most others who have considered the issue remain convinced of the safety of aspartame. At present, the only warning required on products containing it have to do with a danger to phenylketonurics, those individuals unable to metabolize phenylalanine in a normal fashion. As to other possible hazards, the NutraSweet Group of G. D. Searle suggests that a product consumed by 100 million Americans will be associated, quite by chance, with any number of medical, psychological, and social problems.

A few simple calculations tell us that weight will be lost if NutraSweet is substituted for a significant number of sugar-derived calories. For example, if it has been your practice to drink a quart of Coke every day, switching to Diet Coke will cut 360 calories from your daily intake. That translates into a loss of 3 pounds of fat in a month. Why then do we see so many overweight people drinking and eating "diet" this and "diet" that? Shouldn't they all be thin by now?

There are several possible explanations for the paradox of an ever-fatter nation consuming ever-increasing amounts of diet foods. An intriguing suggestion has been made by J. E. Blundell and A. J. Hill, psychologists at Leeds University in England. Their studies in college students indicate that aspartame has appetite-stimulating properties compared with water and that it fails to relieve hunger as does sugar. A still simpler idea is that users of diet drinks, confident that many calories have been saved, go to excess with other foods. Orders for a bacon double cheeseburger and a diet Pepsi fill the air of fast food restaurants. Combinations like that make no sense either calorically or nutritionally. Better to consume the sugar of a regular soft drink than the saturated fat found in bacon, cheese, and beef. Finally, Steven Spellman and Lawrence Garfinkel of the American Cancer Society found in a survey of 79,000 women that the use of artificial sweeteners neither helped weight loss nor prevented weight gain.

If science can provide us with low-calorie sweeteners, cannot science give us synthetic fat? It can and it has. The first to reach the market was Simplesse, a product made from egg whites and casein, a protein found in milk. Approved by the Food and Drug Administration in February 1990, Simplesse is designed to provide the taste and texture of fat, but with only about half the calories and just a touch of real fat. The careers of Simplesse and similar products will surely be interesting; their influence on the stock market is real enough; Monsanto, the parent company of the NutraSweet Division of G. Searle, saw its stock rise $4.38 a share with the FDA approval of Simplesse; their influence upon the weight or health of the nation remains to be seen.

Synthetic foods aside, are there not chemicals that will permit us to proceed directly to our desired state of slimness? There are. Can we not bypass aerobic conditioning, self-discipline, weight training, behavior modification, and similar difficult commitments? Of course we can. Let me illustrate.

During the Vietnam War the son of a Navy captain of my acquaintance brought disgrace upon his father by dodging the draft. He did so in a most modern way without the inconvenience of having to go either to Canada or to jail; by regular ingestion of amphetamines he so depressed his appetite for food that his weight fell below the minimum required to pass the Army physical examination. Today he would likely have used cocaine instead; it would have worked as well.

The medical word for loss of appetite is anorexia. Many drugs, but especially stimulants such as amphetamine and cocaine, are able to suppress appetite. Nicotine, a stimulant of a different type, is likewise an appetite suppressant. Presumably one of the many factors that keep people smoking is the fear that they will gain weight if they stop. The usefulness of anorexogenic agents, as such drugs are called, in weight control is a matter of continuing controversy. What is quite clear, however, is that drug addicts tend to be thin. Use enough cocaine or heroin or alcohol and you will lose weight. Drugs like these not only make you less hungry as a direct effect, they can also substitute for the pleasures of food (in much the same way I suggested you partially substitute the joys of exercise for those of food.) For a variety of reasons, drugs tend to be a very unhealthy way to control weight.

Alright, you've decided that slimness is not worth becoming a coke head. What about all the other pharmacological aids to dieting? These are conveniently classed on the basis of whether or not you need a physician's prescription to get them. Among the prescription drugs are our old friends, the amphetamines, together with a number of other amphetamine-like substances. The latter are sold under tradenames like Apidex-P, Bontril, Preludin, and Tenuate. For each there is concern about drug abuse; in New York State it is illegal for a physician to prescribe amphetamines for weight control.

A single drug, phenylpropanolamine, dominates the non-prescription market. It is found in dozens of brand name products such as Acutrim, Anorexin, Appedrine, Appress, Ayds, Dexatrim, and Prolamine. Pharmacologically, phenylpropanolamine is a pale imitation of amphetamine. In terms of appetite supression, that's bad; it doesn't work as well. On the other hand, phenylpropanolamine has a much diminished abuse liability compared to amphetamine and that's good. Does phenylpropanolamine work? Probably about as well as any of the nonpharmacological gimmicks I mentioned earlier. That is to say, it may help some people some of the time to cut down on caloric intake—wonder drug it is not.

Buried deep in all of us is the thought that someone out there has a potion that will enable us to eat all we want without getting fat. That hope is regularly exploited by various quacks and charlatans. For example, I recently received a:

"MEMO from: Dr. William Byron,
Diet Specialist, Holistic Health Center, London, England."

Dr. Byron kindly tells me that "It's not your fault if you're overweight" and that if I will send just $24.90 to a post office box in Tijuana, Mexico, I can receive a 4-week supply of "Maximum Strength Neutralizer G.H." The promotional materials that accompany this offer would make P. T. Barnum blush. A sample: "Neutralizer G.H. literally pulls excess fat from hard to reach storage sites—waist, hips, thighs, and buttocks—leaving your body thinner, trimmer and sexier than you ever thought possible." And best of all "YOU WILL NOT FAIL THIS TIME. In fact, YOU CANNOT FAIL THIS TIME because you don't have to do anything more than to remember to take your Neutralizer G.H. each and every day."

The general principle here is that if it sounds too good to be true, it probably is, especially if you have to send your money to Tijuana.

Summing Up

According to the Associated Press, Americans spent $33 billion on weight loss in 1989. Think on that a moment: Thirty-three billion dollars spent to eat less. Try as I have, I can think of no greater waste of money. It is a waste not because weight loss is an undesirable goal for many, but because spending money to do it is completely unnecessary. Though little publicized—there's no money in it—a rational plan of diet and exercise along the lines proposed in this chapter has kept millions acceptably shaped for many years.

Guide to Vitamins, Nutrients, Minerals, and Trace Elements

In the brief sections that follow I have summarized much of what is contained in the preceding chapters. I must admit a general lack of enthusiasm for such guides; they are for me similar in appeal to speed reading, sprint drinking, and quick sex. Too often, the statement of isolated "facts" and "myths" gives a false impression of sure knowledge. On the other hand, condensations such as these can serve with little effort to refresh our memories. In addition, they may allay the anxiety created by our daily serving from the media of isolated bits of alarmism.

Macronutrients

Protein

Effects of deficiency: Kwashiorkor, a disease characterized by stunted growth, diminished resistance to infection, mental deterioration, and death.

RDA[a]: Varies with age and body size. Males: 63 grams. Females: 50 grams. During pregnancy and lactation: 60–65 grams.

Sources: Vegetables and grains of all kinds. Meat, fish, poultry, eggs, and milk.

Adverse effects when consumed in excess: Weight gain.

Risk of deficiency: Minimal with a mixed diet. May occur in children, the elderly, and others whose main energy sources are fat and carbohydrate. Vegetable-based infant formulas may not provide adequate protein.

Comments: The consumption of isolated amino acids as food supplements cannot be recommended. In addition to possible hazards such as interference with the absorption of other

---

[a] *Recommended Dietary Allowances*, 10th Edition, National Academy Press, 1989. Values given are for adults of all ages unless otherwise specified.

nutrients, we must be concerned with the source of these amino acids. Tryptophan-induced illness is a recent example.

Fats and Oils

Effects of deficiency: Stunting of growth and development. If other energy sources are provided, only the fatty acids, linoleic and linolenic, appear to be essential. The optimal level of fat in the diet is uncertain.

RDA[a]: None specified.

Sources: Animal: meat, fish, poultry, milk and milk products, eggs, visible fat. Vegetable: margarine, vegetable shortening, oils.

Adverse effects when consumed in excess: Weight gain. Saturated fat may contribute to elevated cholesterol. Possible role of fats, both saturated and unsaturated, in cancer and other diseases is uncertain.

Risk of deficiency: Minimal. Essential fatty acid deficiency has never been observed in normal humans.

Comments: Current recommendations are that fat contribute less than 30% of total calories and that less than 10% be provided by saturated fat.

Carbohydrates and Fiber

Effects of deficiency: Stunting of growth if other energy sources are inadequate. Absence of fiber may contribute to constipation.

RDA[a]: None specified.

Sources: Fruits, vegetables, beans, and grains of all kinds. Refined sugars.

Adverse effects when consumed in excess: Weight gain. Sucrose is the primary cause of tooth decay. Excess fiber, especially in the form of supplements, may cause digestive disturbances including bowel obstruction.

Risk of deficiency: Minimal with a mixed diet.

Comments: Current recommendations are that carbohydrate contribute about 60% of total calories. Preferred sources are fruits, vegetables, beans, and grains. Refined sugars should be consumed in moderation.

Guide to Fat-Soluble Vitamins

Vitamin A

Effects of deficiency: Dryness of the eyes [xerophthalmia], loss of night vision, blindness. Dry skin and mucous membranes. Diminished resistance to infection.

RDA[a]: Male: 1000 RE. female: 800 RE. During breast feeding: 1200–1300 RE. [1 retinol equivalent = 1 microgram of vitamin A or 6 micrograms of beta-carotene]

Sources: [preformed vitamin A] Liver, butter, egg yolk, fortified milk and derived products. [carotene] Green and yellow vegetables, carrots [2025 RE in one medium raw carrot], tomatoes, cantaloupe, other colored fruits and vegetables.

Toxic effects in overdose: Loss of appetite, fatigue, nausea, double vision, loss of hair, dry and peeling skin, bone pain, liver damage.

Risk of deficiency: Most likely in those who consume no milk, fruits, or vegetables.

Comments: Vitamin A in excess is a poison. Beta-carotene seems harmless although you may turn yellow with large excess. If you must supplement, take beta-carotene. Fruits and vegetables high in beta-carotene may protect against certain forms of cancer.

Vitamin D

Effects of deficiency: Inability to absorb calcium from the diet. Rickets in infants and children, osteomalcia in adults.

RDA[a]: Age 6 months to 24 years: 400 IU [an amount = 10 micrograms cholecalciferol]. Pregnant women and nursing mothers: 400 IU. All others: 200 IU.

Sources: Manufactured in human skin exposed to sunlight. Fortified milk and related products, fish oils.

Toxic effects in overdose: Loss of appetite, nausea, vomiting, headache, irritability, fatigue, kidney failure, death.

Risk of deficiency: Most likely in those who are never exposed to sunlight and who consume no fish or dairy products. Examples: housebound elderly, strict vegetarians who practice purdah.

Comments: Vitamin D in excess is a poison.

Vitamin E

Effects of deficiency: Have never been observed in normal persons. In cystic fibrosis and disorders of the digestive system, damage to blood cells and to the nervous system may occur.

RDA[a]: Males: 10 mg alpha-tocopherol equivalents [an amount = approximately 14 IU]. Females: 8 mg [approx. 11 IU]. Pregnant women and nursing mothers: 10–12 mg.

Sources: Widely distributed in eggs, fish, dairy products, fruits, vegetables, and grains.

Toxic effects in overdose: None documented.

Risk of deficiency: Minimal in normal persons.

Comments: Used to protect premature infants against oxygen toxicity. Value of antioxidant properties in slowing the processes of aging is uncertain.

Vitamin K

Effects of deficiency: Bleeding from defective clotting of the blood.

RDA[a]: Males, age 25 and older: 80 micrograms. Females: 65 micrograms. Proportionately less at younger ages.

Sources: Green leafy vegetables, soybeans, eggs, dairy products, fruits, and grains. A significant fraction of the human need is provided by intestinal bacteria that produce vitamin K.

Toxic effects in overdose: None documented. May interfere with the action of anticoagulant drugs.

Risk of deficiency: Minimal in normal adults. Exclusively breast-fed infants are at some risk of deficiency.

Comments: Newborn infants routinely receive vitamin K.

Guide to Water-Soluble Vitamins

Thiamine [Vitamin B$_1$, Thiamin]

Effects of deficiency: Beriberi, a disease characterized by weakness, fatigue, loss of appetite, apathy, depression, constipation, memory loss, disordered function of the heart, nerves, and brain.

RDA[a]: Males, age 11 and over: 1.2–1.5 mg. Females: 1.0–1.1 mg. Pregnant women and nursing mothers: 1.5–1.6 mg.

Sources: Green vegetables, cereals, beans, milk, pork, beef, fortified flour and derived products.

Toxic effects in overdose: None documented.

Risk of deficiency: Greatest in alcoholics due to the combination of a poor diet and diminished absorption.

Comments: Alcoholics are often injected with thiamine to prevent or treat Wernicke-Korsakoff syndrome.

Niacin [Nicotinic Acid]

Effects of deficiency: Pellagra, a disease characterized by weakness, depression, insomnia, and the "3 Ds": dermatitis, diarrhea, and dementia.

RDA[a]: Males, age 11 and over: 15–20 niacin equivalents [1 NE = 1 mg niacin or 60 mg tryptophan]. Females: 13–15 NE. Pregnant women and nursing mothers: 17–20 NE.

Sources: Meat, poultry, dairy products, fortified flour, and derived products.

Toxic effects in overdose: Flushing of the skin, intense itching.

Risk of deficiency: A diet in which corn is the sole source of protein.

Comments: Nicotinamide [niacinamide] can substitute for niacin as a vitamin. Very large doses of niacin may reduce cholesterol; in this use, niacin is a drug; see a physician first.

Riboflavin

Effects of deficiency: Sore throat, inflammation of face, lips, and tongue, anemia.

RDA[a]: Males age 11 and over: 1.5–1.8 mg. Females: 1.2–1.3 mg. Pregnant women and nursing mothers: 1.6–1.8 mg.

Sources: Dairy products, meat, eggs, vegetables, enriched flour, and derived products.

Toxic effects in overdose: None documented.

Risk of deficiency: Minimal with a mixed diet; deficiency common in alcoholics.

Comments: Possible connection with changes in mental state; increased need possible with heavy exercise.

Vitamin B₆ [Pyridoxine]

Effects of deficiency: Convulsions may occur in infants totally deprived of the vitamin.

RDAᵃ: Males age 11 and over: 1.7–2.0 mg. Females: 1.4–1.6 mg. Pregnant women and nursing mothers: 2.1–2.2 mg.

Sources: Whole grains, fruits, vegetables, dried beans, eggs, nuts, meats, poultry.

Toxic effects in overdose: Suppression of breast milk; damage to sensory and motor nerves.

Risk of deficiency: Minimal except with general malnutrition.

Comments: Value in premenstrual syndrome and carpal tunnel syndrome remains to be established.

Biotin

Effects of deficiency: Dry and inflamed skin, loss of hair.

RDAᵃ: "Safe and adequate intake": 30–100 micrograms.

Sources: Milk, eggs, cereals.

Toxic effects in overdose: None documented.

Risk of deficiency: Minimal.

Comments: Raw egg white contains avidin, a protein that binds biotin. Nearly all instances of human biotin deficiency have involved the eating of raw eggs.

Pantothenic Acid

Effects of deficiency: Not clearly defined.

RDAᵃ: "Safe and adequate intake": 4–7 mg.

Sources: Whole grains, dried beans, milk, eggs, fruits and vegetables, meat.

Toxic effects in overdose: None documented.

Risk of deficiency: Minimal.

Comments: "Burning feet" in malnourished prisoners during World War II may have been a specific effect of pantothenic acid deficiency.

Vitamin B₁₂ [Cyanocobalamin]

Effects of deficiency: Anemia, nerve damage, behavioral disturbances.

RDA[a]: 2.0 micrograms. Pregnancy: 2.2 micrograms. Breast feeding: 2.6 micrograms.

Sources: Milk, eggs, meat, fish.

Toxic effects in overdose: None documented.

Risk of deficiency: Greatest in strict vegetarians and their breast-fed
infants.

Comments: Pernicious anemia is a disease in which vitamin B_{12} is
not absorbed from dietary sources. In normal humans, sufficient B_{12} is stored to last for several years.

Folic Acid [Folate, Folacin]

Effects of deficiency: Anemia, behavioral disturbances, complications of pregnancy.

RDA[a]: Males age 15 and over: 200 micrograms. Females: 180
micrograms. Pregnancy: 400 micrograms. Breast feeding:
260–280 micrograms.

Sources: Whole grains, green leafy vegetables, beans.

Toxic effects in overdose: None documented.

Risk of deficiency: Greatest during pregnancy.

Comments: Supplemental folate may mask a deficiency of vitamin B_{12}.

Vitamin C [Ascorbic Acid]

Effects of deficiency: Scurvy, a disease characterized by abnormal
bleeding, defective connective tissue with possible loss of teeth,
weakness, death.

RDA[a]: 60 mg. Pregnancy: 70 mg. Breast feeding: 90–95 mg.

Sources: Citrus fruits, vegetables, and other fruits.

Toxic effects in overdose: No documented toxicity at doses less than
several hundred milligrams per day. Very large doses may
contribute to the formation of kidney stones.

Risk of deficiency: Significant in those who eat no fruits or vegetables;
common in alcoholics.

Comments: Facilitates the absorption of iron from nonmeat sources.
Antioxidant properties may have long-term beneficial effects.

Guide to Minerals

Calcium

Effects of deficiency: Inadequate mineralization of developing bone; weakening of established bone.

RDA[a]: 1200 mg per day until age 25, then 800 mg. 1200 mg during pregnancy and nursing.

Sources: Milk and milk products, bones, dark green leafy vegetables.

Toxic effects in overdose: Constipation, urinary stone formation and kidney damage, interference with the absorption of other minerals.

Risk of deficiency: Overt deficiency is rare. Failure to reach an intake of 1200 mg per day is common whenever milk and milk products are excluded from the diet.

Comments: A quart of skim milk provides 1200 mg of calcium.

Phosphorus

Effects of deficiency: Inadequate mineralization of developing bone in premature infants; weakening of established bone.

RDA[a]: 1200 mg per day until age 25, then 800 mg. 1200 mg during pregnancy and nursing.

Sources: Milk and milk products, meat, fish, and poultry, dried peas and beans, whole grain cereals, soft drinks.

Toxic effects in overdose: May interfere with normal use of calcium.

Risk of deficiency: Minimal.

Comments: One-half pint of skim milk provides 900 mg of phosphorus.

Magnesium

Effects of deficiency: Never observed in normal humans; Diseases in which magnesium is depleted may lead to altered heart and mental functions.

RDA[a]: Maximum of 300–400 mg.

Sources: Whole grain cereals, leafy green vegetables, nuts, soybeans.

Toxic effects in overdose: Fatigue, nausea, vomiting, coma, death.

Risk of deficiency: Minimal in normal persons. Sometimes observed in alcoholics.

Comments: Magnesium supplements are widely promoted. Care must be taken to avoid toxic effects; more is not better.

Iron

Effects of deficiency: Anemia.

RDA[a]: Males: 10–12 mg. Females age 11 to menopause: 15 mg; 30 mg during pregnancy.

Sources: Meat, especially organ meats, egg yolk, peas, beans, nuts, fruits, green leafy vegetables, enriched flour, and cereals.

Toxic effects in overdose: Imbalance in other minerals, liver damage, reduced resistance to infection, impotence.

Risk of deficiency: Most significant during pregnancy and in women and children who eat no animal products.

Comments: Only in the presence of laboratory-documented iron-deficiency anemia and during pregnancy is routine iron supplementation warranted.

Zinc

Effects of deficiency: Loss of appetite, impaired sense of taste and smell, lethargy, stunted growth.

RDA[a]: Males: 15 mg. Females: 12 mg. During pregnancy and while breast feeding: 15–19 mg.

Sources: Milk and milk products, pork , beef, chicken, eggs, legumes.

Toxic effects in overdose: Headache, nausea, cramps, anemia, imbalance in other vitamins and minerals, impaired immune function.

Risk of deficiency: Remains uncertain. An intake less than the RDA may occur with any unbalanced, primarily vegetarian diet but the consequences are largely unknown.

Comments: The presence of a high concentration of zinc in the human prostate has led to coy advertisements that speak of "an active man's good health." Unfortunately there is no evidence that prostatic health is enhanced by excess zinc.

Iodine [Iodide]

Effects of deficiency: Goiter, a condition in which the thyroid gland becomes enlarged. Cretinism, a condition characterized by stunted growth, deafness, and mental retardation.

RDA: 150 micrograms. During pregnancy: 175 micrograms. While breast feeding: 200 micrograms.

Sources: Iodized salt, seafood, milk, vegetables and grains grown in iodide-rich soils.

Toxic effects in overdose: Skin problems, aggravation of acne, disorders of the thyroid gland.

Risk of deficiency: Minimal in the United States because of the use of iodized salt.

Comments: Iodine deficiency continues to be a major problem in South America, India, and Asia.

Selenium

Effects of deficiency: Keshan disease, a heart disorder observed in areas of China in which the soil contains very little selenium.

RDA: Males: 70 micrograms. Females: 55 micrograms. During pregnancy: 65 micrograms. While breast feeding: 75 micrograms.

Sources: Fish, meat, poultry, egg yolk, grains grown in selenium-rich soils.

Toxic effects in overdose: Fatigue, nausea, loss of hair and nails, skin eruptions.

Risk of deficiency: Minimal.

Comments: The promotion of selenium as an "anti-aging," "anti-cancer," and "anti-AIDS" substance and the wide availability of supplements containing as much as 300 micrograms per tablet makes self-poisoning a significant hazard.

Guide to Water and Electrolytes

Water

Effects of deficiency: Disordered kidney and bowel function, dehydration, delerium, and death.

RDA: None established.

Sources: Drinking water, fresh fruits and vegetables, milk, coffee, tea, soft drinks.

Adverse effects when consumed in excess: Usually none. Some commercial weight-loss programs encourage the drinking of several quarts of water each day. Such volumes of water may lead to excessive sodium loss and, in susceptible individuals, to convulsions. [Compulsive water drinking is a feature of some psychotic disorders; coma and, very rarely, death may result.]

Risk of deficiency: Significant in the elderly and in those engaged in strenuous physical activity in hot, dry conditions.

Comments: Despite the claims made for various commercial drinks, pure water is best [and least expensive and lowest in calories].

Sodium

Effects of deficiency: Impaired fluid balance, impaired cellular function, death.

RDA: None established. Minimal intake: 500 mg. Probably safe: 2000–3300 mg.

Sources: Table salt, processed foods, meat, fish, poultry, milk, eggs.

Adverse effects when consumed in excess: Minimal when water intake is adequate. May contribute to high blood pressure in sensitive individuals.

Risk of deficiency: Minimal.

Comments: Three-quarters of the salt in the American diet is added either at the table or in food processing.

Potassium

Effects of deficiency: Impaired fluid balance, impaired cellular function, death.

RDA: None established. Minimal intake: 2000 mg. Probably safe: 1500–6000 mg.

Sources: Fresh fruits and vegetables, oranges, bananas, peanut butter, dried peas and beans, potatoes, yogurt, meat.

Adverse effects when consumed in excess: Minimal when water intake is adequate.

Risk of deficiency: Minimal.

Comments: Potassium deficiency is most often a problem in those persons treated with diuretic agents ["water pills"]; heart irregularities and death may result.

Chloride

Effects of deficiency: Impaired fluid balance, impaired cellular function, death.

RDA[a]: None established. Minimal intake: 750 mg. Probably safe: 1900–5000 mg.

Sources: Table salt, processed foods.

Adverse effects when consumed in excess: Minimal when water intake is adequate.

Risk of deficiency: Minimal.

Comments: Exclusive use of chloride-deficient soy-based formulas has been associated with impaired physical and mental development in infants.

Guide to Trace Elements

Copper

Effects of deficiency: Uncertain in adults. Malnourished children develop anemia and bone defects.

RDA[a]: None established. Estimated safe intake: 1.5–3 mg.

Sources: Shellfish, nuts, beef and pork liver, chocolate, dried beans, raisins, whole grains.

Toxic effects in overdose: Never observed as a result of unsupplemented dietary intake. Excessive supplements may cause anemia, nausea, and vomiting.

Risk of deficiency: Minimal.

Comments: Amounts much in excess of 3 mg may contribute to liver disease and to imbalances in zinc and iron.

Manganese

Effects of deficiency: Never observed in normal humans.

RDA[a]: None established. Estimated safe intake: 2.0–5mg.

Sources: Whole grains, nuts, vegetables, fruits, tea, beets, egg yolks.

Toxic effects in overdose: Never observed as a result of unsupplemented dietary intake.

Risk of deficiency: Minimal.

Comments: Workers exposed to manganese dust or fumes have suffered nerve damage.

Fluoride

Effects of deficiency: Impaired development of bones and teeth.

RDA[a]: None established. Estimated safe intake: 1.5–4 mg.

Sources: Variable amounts are present in all foods and water; tea; fluoridated water and tooth paste.
Toxic effects in overdose: Mottling of teeth, nausea and vomiting.
Risk of deficiency: Minimal where water is fluoridated.
Comments: Although water fluoridation is accepted by most authorities, the subject has remained controversial for half a century. The consequences of a 1990 government report linking fluoride to cancer in rodents are uncertain.

Chromium

Effects of deficiency: Impaired glucose metabolism.
RDA[a]: None established. Estimated safe intake: 50–200 micrograms.
Sources: Fresh fruits and vegetables, meat, cheese, whole grains, dried peas and beans, peanuts.
Toxic effects in overdose: Never observed as a result of unsupplemented dietary intake.
Risk of deficiency: Uncertain.
Comments: Studies over the past decade have suggested a relationship between chromium deficiency and diabetes. The value of chromium supplements remains to be established.

Molybdenum

Effects of deficiency: Never observed in normal humans.
RDA[a]: None established. Estimated safe intake: 75–250 micrograms.
Sources: Milk, peas, beans, cereal grains, organ meats.
Toxic effects in overdose: Interference with copper metabolism.
Risk of deficiency: Minimal.
Comments: Molybdenum supplements are widely available; their value is uncertain.

References

The references listed below for each chapter are far from exhaustive. Instead they are intended to serve as an introduction to the literature of medical science for those with more specific interests. Most of the journals cited will be found in a large general library, health sciences library, or medical library.

For those who wish to obtain expert commentary on current developments in medicine and nutrition, a number of newsletters devoted to aging, fitness, nutrition, and health maintenance are available. I do not pretend to have seen them all, but those from well known institutions, such as Johns Hopkins University, the University of California at Berkeley, and the Mayo Clinic, are of uniformly high quality.

Chapter 1. *Protein: Too much of a good thing?* E. D. Shinwell, R. Gorodischer: Totally vegetarian diets and infant malnutrition. *Pediatrics* **70**: 582–586, 1982. K. J. Carpenter: The history of enthusiasm for protein. *J Nutr* **116**: 1364–1370, 1986. K. Sakimoto: The cause of the eosinophilia–myalgia syndrome associated with tryptophan use. *New Engl J Med* **323**, 992–993 (1990).

Chapter 2. *Fat and essential fatty acids: Lessons from sheep sex.* A. E. Hansen, H. F. Weise, A. N. Boelsche, M. E. Haagard, D. J. D. Adam, H. Davis: Role of linoleic acid in infant nutrition. *Pediatrics* **31**: 171–192, 1963. R. T. Holman, S. B. Johnson,T. F. Hatch: A case of human linolenic acid deficiency involving neurological abnormalities. *Amer J Clin Nutr* **35**: 617–623, 1982. C. H. Hennekens et al.: Preliminary Report: Findings from the aspirin component of the ongoing Physicians' Health Study. *New Engl J Med* **318**: 262–264, 1988. J. L. Burton: Dietary fatty acids and inflammatory skin disease. *Lancet*, January 7, 1989, 27–31. R. A. Lewis, K. F. Austen, R. J. Soberman: Leukotirenes and other products of the 5-lipoxygenase pathway. Biochemistry and relation to pathobiology and human disease. *New Engl J Med* **323**: 645–655, 1990. R. P. Mensink and M. B. Katan: Effect of *trans* fatty acids on high-density and low-density lipoprotein cholesterol levels in healthy subjects. *New Engl J. Med* **323**: 439–445, 1990.

Chapter 3. *Carbohydrates: From glucose to oat bran to sawdust.* R. A. McCance, E. M. Widdowson: Old thoughts and new work on breads white and brown. *Lancet* July 30, 1955: 205–210. D. P. Burkitt, A. R. P. Walker, N. S. Painter: Dietary fiber and disease. *JAMA* **229**: 1068–1074, 1974. G. L. Simon, S. L. Gorbach: Intestinal flora in health and disease. *Gastroenterology* **86**: 174–193, 1984. D. Kritchevsky: Dietary fiber. *Ann Rev Nutr* **8**: 301–328, 1988. W. E. Connor: Dietary fiber—Nostrum or critical nutrient? *New Engl J Med* **322**: 439–445, 1990.

Chapter 4. *Is sugar a poison?* P. A. Crapo, J. M. Olefsky: Food fallacies and blood sugar. *N Engl J Med* **309**: 44–45, 1983. M. D. Gross: Effect of sucrose on hyperkinetic children. *Pediatrics* **74**: 876–878, 1984. House of Delegates: Position paper of the American Dietetic Association on diet and criminal behavior. *J Am Dietetic Assoc* **85**: 361–362, 1985. J. H. Shaw: Causes and control of dental caries. *N Engl J Med* **317**: 996–1004, 1987. J. P. Bantle: The dietary treatment of diabetes mellitus. *Med Clin NA* **72**: 1285–1299, 1988. R. L. Nelson: Oral glucose tolerance test: Indications and limitations. *Mayo Clin Proc* **63**: 263–269, 1988. J. Bachorowski, J. P. Newman, S. L. Nichols, D. A. Gans, A. E. Harper, S. L. Taylor: Sucrose and delinquency. *Pediatrics* **86**: 245–253, 254–262, 1990. H. King and J. E. Dowd: Primary prevention of type 2 [non-insulin-dependent] diabetes mellitus. *Diabetologia* **33**: 3–8, 1990.

Chapter 5. *Vitamin A and beta-carotene: Of carrots and toxins.* H. W. Josephs: Hypervitaminosis A and carotenemia. *Am J Dis Child* **67**: 33–43, 1944. A. Gerber, A. P. Raab, A. E. Sobel: Vitamin A poisoning in adults. *Am J Med* **16**: 729–745, 1954. J. A. Olson: Recommended dietary intakes of vitamin A in humans. *Am J Clin Nutr* **45**: 704–716, 1987. A. Bendich, J. A. Olson: Biological actions of carotenoids. *FASEB J* **3**: 1927–1932, 1989. G. T. Keusch: Vitamin A supplements—Too good not to be true. *New Engl J Med* **323**: 985–987, 1990. J. N. Hathcock, D. G. Hatton, M. Y. Jenkins, J. T. McDonald, P. R. Sundaresen, V. L. Wilkening: Evaluation of vitamin A toxicity. *Am J Clin Nutr* **52**: 183–202, 1990.

Chapter 6. *Thiamine (Vitamin B₁): Polished rice and beriberi.* M. K. Horwitt: Interpretations of requirements for thiamin, riboflavin, niacin-tryptophan, and vitamin E plus comments on balance studies and vitamin B₆. *Am J Clin Nutr* **44**: 973–985, 1986. R. H. Haas: Thiamin and the brain. *Ann Rev Nutr* **8**, 483–515, 1988. C. M. Tang, M. Rolfe, J. C. Wells, K. Cham: Outbreak of beri-beri in the Gambia. *Lancet*, July 22, 1989, 206–207. A. Wodak, R. Richmond, A. Wilson: Thiamin fortification and alcohol. *Med J Australia* **152**: 97–99, 1990.

Chapter 7. *Niacin: Pellagra and madness.* J. Goldberger, W. F. Tanner: Amino-acid deficiency probably the primary etiological factor in pellagra. *Public Health Rep* **37**: 462–486, 1922. S. Eichold, M. Velek: Pellagra: A revisionist approach. *Amer J Med* **85**: 405–406, 1988. S. C. Litin and C. F. Anderson: Nicotinic acid-associated myopathy: A report of three cases. *Am J Med* **86**: 481–483, 1989. A. Garg and S. M. Grundy: Nicotinic acid as therapy for dyslipidemia in non-insulin-dependent diabetes mellitus. *JAMA* **264**: 723–726, 1990.

Chapter 8. *Riboflavin: A vitamin for depression?* A. Z. Belko, M. P. Meredith, H. J. Kalkwarf, E. Obarzanek, S. Weinberg, R. Roach, G. McKeon, D. A. Roe: Effects of exercise on riboflavin requirements: biological validation in weight reducing women. *Am J Clin Nutr* **41**: 270–277, 1985. D. B. McCormick: Two interconnected B vitamins: Riboflavin and pyridoxine. *Physiol Rev* **69**: 1170–1198, 1989. K. Suboticanec, A. Stavljenic, W. Schalch, R. Buzina: Effects of pyridoxine and riboflavin supplementation on physical fifitness in young adolescents. *Internat J Vit Nutr Res* **60**: 81–88, 1990.

Chapter 9. *Vitamin B₆: Morning sickness and the premenstrual syndrome.* H. C. Hesseltine: Pyridoxine failure in nausea and vomiting of pregnancy. *Am J Ob Gyn* **51**: 82–86, 1946. S. E. Snyderman, L. E. Holt, R. Carretero, K. Jacobs: Pyridoxine deficiency in the human infant. *J Clin Nutr* **1**: 200–202, 1953. C. J. Molony, A. H. Parmalee: Convulsions in young infants as a result of pyridoxine deficiency. *JAMA* **154**: 405–408, 1954. D. Gath and S. Iles: Treating the premenstrual syndrome. *BMJ* **297**: 237–238, 1988. K. E. Kendall and P. P. Schnurr: The effects of vitamin B₆ supplementation on premenstrual symptoms. *Obstet Gynecol* **70**: 45–149. 1987.

Chapter 10. *Pantothenic acid: The anti-gray hair vitamin.* B. Lund, H. Kringstad: The anti grey hair vitamin, a new factor in the vitamin B complex. *J Nutr* **19**: 321–331, 1940. C. Gopalan: The "burning feet" syndrome. *Indian Med Gaz* **46**: 22–25, 1946. M. Glusman: The syndrome of "burning feet" (nutritional melalgia) as a manifestation of nutritional deficiency. *Am J Med* **3**: 211–223, 1947. R. E. Hodges, M. A. Olson, W. B. Bean: Pantothenic acid deficiency in man. *J Clin Invest* **37**: 1642–1657, 1958. B. R. Eissenstat, B. W. Wyse, R. G. Hansen: Pantothenic acid status of adolescents. *Am J Clin Nutr* **44**: 931–937, 1986.

Chapter 11. *Biotin: Did Rocky know about avidin?* M. A. Boas: The effect of desiccation upon the nutritive properties of egg-white. *Biochem J* **21**: 712–724, 1927. V. P. Sydenstricker, S. A. Singal, A. P. Briggs, N. M. DeVaughn,H. Isbell: Observations on the "egg white injury" in man. *JAMA* **118**: 1199–1200, 1942. R. H. Williams: Clinical biotin deficiency. *N Engl J Med* **228**: 247–252, 1943. K. A. Lombard and D. M. Mock: Biotin nutritional status of vegans, lactoovovegetarians, and nonvegetarians. *Am J Clin Nutr* **50**: 486–490, 1989. B. Wolf and G. S. Heard: Screening for biotinidase deficiency in newborns: Worldwide experience. *Pediatrics* **85**: 512–517, 1990.

Chapter 12. *Vitamin B₁₂: Dr. Castle's predigested hamburger.* G. R. Minot, W. P. Murphy: Treatment of pernicious anemia by a special diet. *JAMA* **250**: 3328–3335, 1926. V. Herbert: Recommended dietary intakes of vitamin B₁₂ in humans. *Am J Clin Nutr* **45**: 671–678, 1987. L. Lawhorne, D. Ringdahl: Cyanocobalamine injections for patients without documented deficiency.*JAMA* **261**: 1920–1923, 1989. J. E. Kaslow, L. Rucker, R. Onishi: Liver extract—folic acid—cyanocobalamin vs placebo for chronic fatigue syndrome. *Arch Int Med* **14**: 2501–2503, 1989. P. Bar-Sella, Y. Rakover, D. Ratner: Vitamin B₁₂ and folate levels in long-term vegans. *Isr J Med Sci* **26**: 309–312, 1990.

Chapter 13. *Folic Acid: Lucy Wills in the slums of Bombay.* L. Wills, B. D. F. Evans: Tropical macrocytic anemia: Its relation to pernicious anemia. *Lancet*, August 20, 1938, 416–421. R. W. Heinle, A. D. Welch: Folic acid in pernicious anemia. Failure to prevent neurological relapse. *JAMA* **133**: 739–741,1947. V. Herbert: Experimental nutritional folate deficiency in man. *Trans Assoc Amer Phys* **75**: 307–320, 1962. V. Herbert: Recommended dietary intakes of folate in humans. *Am J Clin Nutr* **45**: 661–670, 1987. A. F. Subar, G. Block, L. D. James: Folate intake and food sources in the US population. *Am J Clin Nutr* **50**: 508–516, 1989. C. E. Butterworth and T. Tamura: Folic acid safety and toxicity: A brief review. *Am J Clin Nutr* **50**: 53–58, 1989.

Chapter 14. *Vitamin C: Scurvy, Dr. Pauling, cancer, and colds.* J. H. Crandon, C. C. Lund, D. B. Dill: Experimental human scurvy. *N Engl J Med* **223**: 353–368, 1940. T. W. Anderson, G. H. Beaton, P. N. Corey, L. Spero: Winter illness and vitamin C: The effect of relatively low doses. *CMAJ* **112**: 823–826, 1975. E. Cameron, L. Pauling: Supplemental ascorbate in the supportive treatment of cancer. I. Prolongation of survival times in terminal human cancer. *Proc Natl Acad Sci USA* **73**: 3685–3689, 1976. E. T. Creagan, C. G. Moertel, J. R. O'Fallon, A. J. Schutt, M. J. O'Connell, J. Rubin, S. Frytak: Failure of high-dose vitamin C (ascorbic acid) therapy to benefit patients with advanced cancer. *N Engl J Med* **301**: 687– 690, 1979. C. G. Moertel, T. R. Fleming, E. T. Creagan, J. Rubin, M. J. O'Connell, M. A. Ames: High-dose vitamin C versus placebo in the treatment of patients with advanced cancer who have had no prior chemotherapy. *N Engl J Med* **312**: 137–141, 1985. M. Levine: New concepts in the biology and biochemistry of ascorbic acid. *N Engl J Med* **314**: 892–902, 1986. J. A. Olson, R. E. Hodges: Recommended dietary intakes (RDI) of vitamin C in humans. *Am J Clin Nutr* **45**: 693–703, 1987.

Chapter 15. *Vitamin D: Hormone or vitamin?* E. V. McCollum, N. Simmonds, J. Ernestine: Studies on experimental rickets. XXI. An experimental demonstration of the existence of a vitamin which promotes calcium deposition. *J Biol Chem* **53**: 293–312, 1922. M. S. Seelig: Vitamin D and cardiovascular, renal, and brain damage in infancy and childhood. *Ann NY Acad Sci* **147**: 537–582, 1969. I. Hayward, M. T. Stein, M. I. Gibson: Nutritional rickets in San Diego. *Am J Dis Child* **141**: 1060–1062, 1987. M. S. Schwartzman, W. A. Franck: Vitamin D toxicity complicating the treatment of senile, postmenopausal, and glucocorticoid-induced osteoporosis. *Am J Med* **82**: 224–230, 1987. H. Reichel, H. P. Koeffler, A. W. Norman: The role of the vitamin D endocrine system in health and disease. *N Engl J Med* **320**: 980–991, 1989. R. W. Chesney: Requirements and upper limits of vitamin D intake in the term neonate, infant, and older child. *J Pediat* **116**: 159–166, 1990.

Chapter 16. *Vitamin E: A vitamin in search of a disease.* T. W. Anderson: Vitamin E in angina pectoris. *CMAJ* **110**: 401–406, 1974. I. D. Raacke: Herbert McLean Evans (1882–1971), A biographical sketch. *J Nutr* **5**: 929–943, 1983. J. G. Bieri, L. Corash, V. S. Hubbard: Medical uses of vitamin E. *N Engl J Med* **308** 1063–1071, 1983. A. Bendich, L. J. Machlin: Safety of oral intake of vitamin E. *Am J Clin Nutr* **48**: 612–619, 1988. M. K. Horwitt: Supplementation with vitamin E. *Am J Clin J Nutr* **47**: 1088–1089, 1988. L. J. Howard: Laboratory and electrophysiological assessment. *Nutr Rev* **48**: 169–177, 1990.

Chapter 17. *Calcium: Dietary Factor of the Year.* D. M. Hegsted, I. Moscoso, C. Callazos: A study of the minimum calcium requirements of adult men. *J Nutr* **46**: 181–201, 1952. A. R. P. Walker: The human requirement of calcium: Should low intakes be supplemented? *Am J Clin Nutr* **25**: 518–530, 1972. J. A. Kanas and R. Passmore: Calcium supplementation of the diet: Not justified by present evidence. *Brit Med J* **298**: 137–140, 205–208, 1989. B. E. C. Nordin and R. P. Heaney: Calcium supplementation of the diet: Justified by present evidence. *Brit Med J* **300**: 1056–1060, 1990.

Chapter 18. *Iron: Tonic or toxin?* V. Herbert: Recommended dietary intakes [RDI] of iron in humans. *Am J Clin Nutr* 45: 679–686, 1987. C. K. Arthur, J. P. Isbister: Iron deficiency. Misunderstood, misdiagnosed, mistreated. *Drugs* 33: 171–182, 1987. C. Hershko, T. Peto, D. J. Weatherall: Iron and infection. *Brit Med J* 296: 660–664, 1988. R. G. Stevens, D. Y. Jones, M. S. Micozzi, P. R. Taylor: Body iron stores and risk of cancer. *N Engl J Med* 319: 1047–1052, 1988. I. J. Newhouse, D. B. Clement, J. E. Taunton, D. C. McKenzie: The effects of prelatent/latent iron defificiency on physical work capacity. *Med Sci Sports Exerc* 21: 263–268, 1989. N. D. C. Finlayson: Hereditary [primary] haemochromatosis. A common condition in which early recognition may be life saving. *Brit Med J* 301: 350–351, 1990. R. D. Baynes and T. H. Bothwell: Iron deficiency. *Ann Rev Nutr* 10: 133–148, 1990.

Chapter 19. *The Training Effect.* J. M. Rippe, A. Ward, J. P. Porcari, P. S. Freedson: Walking for health and fitness. *JAMA* 259: 2720–2724, 1988. K. E. Powell, C. J. Casperson, J. P. Koplan, E. S. Ford: Physical activity and chronic diseases. *Am J Clin Nutr* 49: 999–1006, 1989. M. A. Fiatarone, E. C. Marks, N. D. Ryan, C. N. Meredity, L. A. Lipsitz, W. J. Evans: High-intensity strength training in nonagenarians. *JAMA* 263: 3029–3034, 1990. J. N. Morris, D. G. Clayton, M. G. Everitt, A. M. Semmence, E. H. Burgess: Exercise in leisure time: coronary attack and death rates. *Brit Heart J* 63: 325–334, 1990.

Chapter 20. *A Program of Exercise.* D. S. Ballor, V. L. Katch, M. D. Becque, C. R. Marks: Resistance weight training during caloric restriction enhances lean body weight maintenance. *Am J Clin Nutr* 47: 19–25, 1988. J. A. Blumenthal et al.: Cardiovascular and behavioral effects of aerobic exercise training in healthy older men and women. *J Gerentol Med Sci* 44: M147–157, 1989. V. E. Friedewald, D. W. Spence: Sudden cardiac death associated with exercise: The risk–benefit issue. *Am J Cardiol* 66: 183–188, 1990. S. N. Blair, H. W. Kohl, R. S. Paffenbarger, D. G. Clark, K. H. Cooper, L. W. Gibbons: Physical fitness and all-cause mortality. *JAMA* 262: 2395–2401, 1990.

Chapter 21. *Cancer: Can 90% be Prevented?* J. C. Bailar III, E. M. Smith: Progress against cancer? *N Engl J Med* 314: 1226–1232, 1986. *Am J Clin Nutr* 43: 629–635, 1986. J. Higginson: Changing concepts in cancer prevention: Limitations and implications for future research in environmental carcinogenesis. *Cancer Res* 48: 1381–1389, 1988. R. Doll: Epidemiology and the prevention of cancer: Some recent developments. *J Cancer Res Clin Oncol* 114: 447–458, 1988. J. L. Freudenheim and S. Graham: Toward a dietary prevention of cancer. *Epidemiol Rev* 11: 229–235, 1989. G. R. Howe et al.: Dietary factors and risk of breast cancer: Combined analysis of 12 case-control studies. *J Natl Cancer Inst* 82: 561–569, 1990. D. L. Davis, D. Hoel, J. Fox, A. Lopez: International trends in cancer mortality in France, West Germany, Italy, Japan, England and Wales, and the USA. *Lancet*, August 25, 1990, 474–497.

Chapter 22. *Heart Attack and Stroke: The basics.* O. Paul et al.: Multiple Risk Factor Intervention Trial. *JAMA* 248: 1465–1477, 1982. B. S. Rifkind et al.: The Lipid Research Clinics Coronary Primary Prevention Trial Results. *JAMA* 251: 351–375, 1984. J. W. Anderson: Dietary fiber, lipids, and atherosclerosis.

Am J Cardiol **60**: 17G-22G, 1987. A. Nordoy and S. H. Goodnight: Dietary lipids and thrombosis. *Arteriosclerosis* **10**: 149-163, 1990. S. M. Grundy, D. S. Goodman, B. M. Rifkind, J. I. Cleeman: The place of HDL in cholesterol management. *Arch Intern Med* **149**: 505-510, 1990. D. M. Dreon, K. M. Vranizan, R. M. Krauss, M. A. Austin, P. D. Wood: The effects of polyunsaturated fat vs monounsaturated fat on plasma lipoproteins. *JAMA* **263**: 2462-2466, 1990.

Chapter 23. *Heart Attack and Stroke: The role of diet and exercise.* J. N. Morris, J. A. Heady, P. A. B. Raffle, C. G. Roberts, J. W. Parks: Coronary heart disease and physical activity at work. *Lancet* November 21, 1953, 1053-1057; November 28, 1953, 1071-1077. H. M. Sinclair: Deficiency of essential fatty acids and atherosclerosis. *Lancet* **1**: 381, 1956. G. J. Miller, N. E. Miller: Plasma-high-density-lipoprotein concentration and development of ischemic heart disease. *Lancet*, Jan 4, 1975, 16-19. D. Kromhout, E. B. Bosscheiter, C. DeL. Coulander: The inverse relation between fish consumption and 20-year mortality from coronary heart disease. *N Engl J Med* **312**: 1205-1209, 1985. H. R. Knapp, G. A. FitzGerald: The antihypertensive effects of fish oil. *N Engl J Med* **320**: 1037-1043, 1989. R. Stamler et al.: Primary prevention of hypertension by nutritional-hygienic means. *JAMA* **262**: 1801-1807, 1989. D. Ornish et al.: Can lifestyle changes reverse coronary heart disease? *Lancet* **336**: 129-133, 1990. M. F. Muldoon, S. B. Manuck, K. A. Matthews: Lowering cholesterol concentrations and mortality: A quantitative review of primary prevention trials. *Brit Med J* **301**: 309-314, 1990.

Chapter 24. *The Prevention of Osteoporosis.* V. Matkovic, K. Kostial, I. Simonovic, R. Buzina, A. Brodarec, B. E. C. Nordin: Bone status and fracture rates in two regions of Yugoslavia. *Am J Clin Nutr* **32**: 540-549, 1979. E. L. Smith, C. Gilligan, M. McAdam, C. P. Ensign, P. E. Smith: Deterring bone loss by exercise intervention in premenopausal and postmenopausal women. *Calcif Tissue Int* **44**: 312-321, 1989. P. B. Gleeson, E. J. Protas, A. D. LeBlanc, V. S. Schneider, H. J. Evans: Effects of weight lifting on bone mineral density in premenopausal women. *J Bone Mineral Res* **5**: 153-158, 1990. R. Sitruk-Ware: Estrogen therapy during menopause. Practical treatment recommendations. *Drugs* **39**: 203-217, 1990. B. Dawson-Hughes, G. E. Dallal, E. A. Krall, L. Sadowski, A. Sahyoun, S. Tannenbaum: A controlled trial of the effect of calcium supplementation on bone density in postmenopausal women. *N Engl J Med* **323**: 878-883, 1990.

Chapter 25. *The Beverly Hills Diet and other tales of the supernatural.* J. Hirsch, R. L. Liebel: New light on obesity. *N Engl J Med* **318**: 509-510, 1988. Council on Scientific Affairs: Treatment of obesity in adults. *JAMA* **260**: 2547-2551, 1988. R. A. Gelfand, R. Hendler: Effect of nutrient composition on the metabolic response to very low calorie diets: Learning more and more about less and less. *Diabet/Metab Rev* **5**: 17-30, 1988. A. N. Howard: The historical development of very low calorie diets. *Int J Obesity* **13** [S-2]: 1-9, 1989. A. Keys: Longevity of man: Relative weight and fatness in middle age. *Annals Med* **21**: 163-168, 1989. E. A. H. Sims: Destiny rides again as twins overeat. *N. Engl J Med* **322**: 1522-1524, 1990. T. A. Wadden, T. B. VanItallie, G. L. Blackburn: Responsible and irresponsible use of very-low-calorie diets in the treatment of obesity. *JAMA* **263**: 83-85, 1990.

Index